VOID

Library of
Davidson College

OXFORD LOGIC GUIDES

GENERAL EDITOR: DANA SCOTT

CLASSICAL PROPOSITIONAL OPERATORS

AN EXERCISE IN THE FOUNDATIONS OF LOGIC

By
KRISTER SEGERBERG
Professor of philosophy
University of Auckland

CLARENDON PRESS · OXFORD
1982

Oxford University Press, Walton Street, Oxford OX2 6DP
London Glasgow New York Toronto
Delhi Bombay Calcutta Madras Karachi
Kuala Lumpur Singapore Hong Kong Tokyo
Nairobi Dar es Salaam Cape Town
Melbourne Auckland
and associates in
Beirut Berlin Ibadan Mexico City Nicosia

© Krister Segerberg 1982

Published in the United States
by Oxford University Press, New York

All rights reserved. No part of this publication may be reproduced, stored in a retrieval system, or transmitted, in any form or by any means, electronic, mechanical, photocopying, recording, or otherwise, without the prior permission of Oxford University Press

British Library Cataloguing in Publication Data
Segerberg, Krister
   Classical propositional operators.
   — (Oxford logic guides).
   1. Propositional calculus
   I. Title
   511.3   QA9.3   80–41798
   ISBN 0–19–853173–7

Typeset by Anne Joshua Associates, Oxford
Printed in Great Britain
at the University Press, Oxford
by Eric Buckley
Printer to the University

TO FINLAND

PREFACE

This is a book on abstract propositional logic, but it began as a book on modal logic. The first question confronting anyone setting out to write a book on modal logic nowadays is, or ought to be, what sort of thing a modal logic is. An introductory chapter of the projected book was accordingly set aside in order to dispose of this question. As I was to realize, it is not as simple as it may appear — at least not if more than a merely stipulative answer is required and one that yields insight. The would-be chapter grew out of proportion, and in the end I decided to publish the eventual jumbo chapter on its own. In preparing the final text I have tried to de-emphasize the modal connection, but it is there, and in Chapter 6 it is explained in detail.

No doubt it is a defect of this book that I have not done more to trace the history of the abstract study of propositional logic (which goes back at least to Tarski) and to relate this study to the present state-of-affairs of that discipline. However, I have received almost all of my own inspiration from Dana Scott and his writings (see the bibliography) and I have felt it best to leave the writing of history to others. I also wish to acknowledge an important impulse provided by David Makinson which is explained in Chapter 4.

An unsuccessful first draft of the book was attempted at Ojmundsbods Fiskeläge during the summer of 1975, but the first complete draft was written in 1977 during an extended visit to the University of Kansas. The result was criticized by Steven T. Kuhn, now of Georgetown University, and David Makinson of UNESCO and the American University of Beirut — extensively, scathingly, but constructively. The final version was read by Dana Scott, editor of the series in which book now appears. I am enormously indebted to these three for all the time they spent on my difficult typescript, and to Rex Martin, then chairman of the philosophy department at K.U., for the excellent working conditions I found there. The material of the book has served as a basis for a seminar in the University of Kansas and a lecture series in Åbo Academy, and I gratefully acknowledge the feedback from the students who took these courses, especially Richard Fleming of Lawrence and Patrick Sibelius and Jari Talja of Åbo/Turku.

The bulk of the final manuscript was completed in November 1979. Sections 0.3 and 3.7 were added in October 1980, and Chapter 6 in May 1981. A good deal of the work was done during two half-year fellowships (the spring terms of

1977 and 1979) awarded by the Academy of Finland (Statens Humanistiska Kommission), and I should like to record my thanks for this as well as for much other support over the years. I should also like to take this opportunity to express my general affection for Risto Hilpinen, Jussi Pietarinen, and Arto T. Salomaa, long-time colleagues at the other university of Turku, Turun Yliopisto.

I dedicate this book to Finland, my ex-step-country, where for ten years my family and I had our home.

Auckland K.S.
Ascension Day 1981

CONTENTS

Preface	vii
0. Induction: some remarks	1
0.1. Induction over the set of natural numbers	1
0.2. Induction over other structures	5
0.3. Tree-structures and trees	12
1. Propositional languages	22
1.1. Propositional languages	22
1.2. Expressions	23
1.3. Formulas	25
1.4. Subformulas	28
1.5. Substitution	31
2. Common logics	34
2.1. Logics	34
2.2. Some common conditions on logics	36
2.3. Further conditions on logics	38
2.4. Syntactic equivalence	42
2.5. Extensions	44
2.6. The lemmas of Tukey and Lindenbaum	47
3. Boolean logics	51
3.1. Boolean operators	51
3.2. Boolean semantics	52
3.3. Types of Boolean operators	56
3.4. A completeness theorem in Boolean logic	59
3.5. Congruence in Boolean logic	63
3.6. Particular truth-value functions: the Big Seven	66
3.7. Arbitrary truth-value functions	73
4. Pre-classical logics	85
4.1. Expressibility	85
4.2. Boolean extensions	88
4.3. A conservation theorem in pre-classical logic	92
4.4. The Big Seven again	94

4.5. The Deduction Theorem and the Finiteness Theorem in pre-classical logic	98
4.6. Makinson's Warning	100
5. Classical logics	106
5.1. Replacement of tautological equivalents: formulas	106
5.2. Replacement of tautological equivalents: sets of formulas	108
5.3. Classical operators	110
5.4. The lattice of classical logics	113
5.5. The Isomorphism Theorem	115
6. Modal logics: a postscript	125
6.1. Two ambitions	125
6.2. A question in modal logic	126
6.3. Some translations between particular modal languages	130
6.4. Two theorems on translations	134
6.5. A definition of modal logic	138
References	143
Index of terms	145
Index of symbols	150
Index of names	152

0
INDUCTION: SOME REMARKS

0.1. Induction over the set of natural numbers

To say that induction is important in mathematical logic is an understatement: it is indispensable, and in this book it is used constantly. But students unfamiliar with induction are sometimes puzzled. It is for the benefit of such students that this introductory chapter has been written.

Nothing in later chapters hinges on this first one, occasional references notwithstanding, except to the extent that the subject matter is assumed to be known. Therefore experienced readers may — perhaps should — skip the entire chapter. Also inexperienced readers may skip this chapter on a first reading and go back to it only if the need arises.

Let us begin with an arithmetical example. How much is $0 + 1 + \ldots + n$? There is a well-known notation for this quantity,

(†) $$\sum_{i=0}^{n} i.$$

But what we are asking for is, of course, an evaluation of this sum — some simple expression in terms of n and the usual arithmetical operations.

It is worth noticing that (†) may be regarded as a function from the set $\omega = \{0, 1, 2, \ldots\}$ of natural numbers into itself. This would perhaps be more evident if in place of the traditional notation we would use s(n) for the sum of the natural numbers up to and including n. In fact, this function is characterized by the following conditions:

$$s(0) = 0,$$
$$s(n + 1) = s(n) + (n + 1),$$

where n is a parameter ranging over ω (which means that the latter condition holds for every value of n). The answer to our question can then be given and proved as follows.

CONTENTION. *For every n, the following equation holds:*

$$(S_n) \qquad\qquad s(n) = \frac{n(n+1)}{2}.$$

Proof. Notice that for distinct m and n, (S_m) and (S_n) are two different statements. Thus there exists indefinitely many statements of this sort; our task is to prove that they are all true.

The proof consists of two steps.

Basic step. We note that (S_0) is true. This is easy to see: $s(0) = 0$ by definition, and $0(0+1)/2 = 0$ by calculation.

Inductive step. Let n be any particular natural number. We assume — as our *induction hypothesis*, as we shall call it — that (S_n) is true. In fact, the induction hypothesis is so important that it may be worth displaying it:

($\S$) $\qquad\qquad\qquad (S_n)$ is true.

From this it follows that also (S_{n+1}) is true:

$$s(n+1) = s(n) + (n+1), \text{ by the definition of s,}$$

$$= \frac{n(n+1)}{2} + (n+1), \text{ by the induction hypothesis, ($\S$),}$$

$$= \frac{(n+1)((n+1)+1)}{2}, \text{ by elementary reasoning.}$$

This completes the proof. ∎

The preceding is a simple example of a proof 'by induction'. It is interesting to ask just why it is a proof. An argument might run like this. The contention to be proved is equivalent to the claim that, for every n, (S_n) is true. This in turn is the same as the claim that $(S_0), (S_1), (S_2),\ldots$ all are true. This is what our proof accomplishes. For the inductive step guarantees that the following infinitely many implications all hold:

if (S_0) then (S_1)
if (S_1) then (S_2)
if (S_2) then (S_3)
$\vdots$

All we need, then, is the truth of (S_0). And that is supplied by the basic step.

All this is 'readily seen', to resort to a cover-up term commonly used for the obvious or the tedious or — sometimes — the obscure. What is problematic in our example is that it is humanly impossible to give infinitely many individual proofs. What saves us is that the desired proofs are so similar. So similar, in fact, that they can be formulated in the same way and so become identical. The inductive

INDUCTION: SOME REMARKS 3

step takes care of infinitely many individual proofs in one blow, so to speak.

What we are tacitly relying on in the proof of the contention above is the principle of induction. Some have suggested that it is a self-evident principle. Poincaré regarded it as the prime example of a synthetic *a priori* truth. Others have felt that it is one of the defining properties of the set of natural numbers. But everyone agrees that it is true.

To bring out this principle it is helpful to have another look at the preceding example. Let us say of a natural number n that it has the property Q if it satisfies the condition

$$s(n) = \frac{n(n+1)}{2}.$$

This certainly defines a particular numerical property. Now in our proof the basic step establishes that 0 has the property Q (which we may write Q(0) or just Q0). The inductive step shows that Q is 'inherited' by larger numbers in the sense that whenever a natural number has Q, then so does its immediate successor. Therefore, it is understood, all numbers have the property Q.

What is needed to support the last 'therefore' is the *principle of mathematical induction*. It may be formulated as follows: for every property P,

$$P(0) \,\&\, \forall n(P(n) \Rightarrow P(n+1)) \Rightarrow \forall n P(n).$$

If one uses the idiom of set theory rather than that of logic, the same principle is rendered by saying that, for every subset $S \subseteq \omega$,

$$0 \in S \,\&\, \forall n(n \in S \Rightarrow n+1 \in S) \Rightarrow S = \omega.$$

(In both cases the quantifiers are assumed to range over ω.)

Three remarks are in order at this point. One concerns terminology: we will not always use the expression 'mathematical induction'. Among the synonyms will be '(ordinary) induction', 'induction over ω' and 'induction on n' (to indicate the *induction variable*, as it is called).

The second remark is of theoretical interest. Our account of induction has been informal. In particular, our use of logical notation has been for succinctness and as part of the metalanguage; this is how it is normally used in this book. But in a more rigorous treatment, one would wish to formalize the principle of induction within a formal language. A particular interest would then accrue to the phrase 'for every property P' (respectively, the phrase 'for every subset $S \subseteq \omega$'). If one uses second-order predicate logic, the induction principle can be expressed by one formula:

$$\forall P(P(0) \,\&\, \forall n(P(n) \Rightarrow P(n+1)) \Rightarrow \forall n P(n)).$$

Within first-order logic so straightforward a formalization is not available. However, here one may proceed in a more roundabout way. Let $P_0(n), P_1(n),\ldots, P_k(n),\ldots$ be an enumeration of all formulas in which n is the only free variable.

The induction principle is then expressed by a denumerably infinite family $\{I_k : k \in \omega\}$ of first-order formulas, of which each I_k may be regarded as an instance of the induction principle:

$$(I_k) \qquad P_k(0) \,\&\, \forall n(P_k(n) \Rightarrow P_k(n+1)) \Rightarrow \forall n P_k(n).$$

The two formulations are not equivalent, the former being the stronger one. However, nothing in our presentation will depend on this distinction, and so we will ignore it here.

The third remark is methodological and perhaps the most important one. Throughout the book we maintain a sharp distinction between object level and metalevel. On the object level are the things that we study: formal languages, formal logics, and other objects to be defined. On the metalevel are we who reason about these objects. And in our reasoning we take a good many things for granted: standard modes of argument as well as so-called naïve set theory. In particular we take the natural numbers for granted, and we regard the principle of induction over them as a correct principle without offering any proof of its correctness. It is important to realize this: we try to explain induction and help the reader understand it (however 'understand' is to be analysed), but we make no effort to derive it from more basic principles or to provide a deeper foundation for it.

However, there are other induction principles which we do not take for granted but which can be derived from the basic principle of mathematical induction. One of them is the *principle of course-of-values induction*, to which we now turn. It may be formulated as follows: for every property P,

$$\forall n(\forall i(i < n \Rightarrow P(i)) \Rightarrow P(n)) \Rightarrow \forall n P(n).$$

Or, equivalently: for every subset $S \subseteq \omega$,

$$\forall n(\forall i(i < n \Rightarrow i \in S) \Rightarrow n \in S) \Rightarrow S = \omega.$$

From these formulations it may not be immediately clear how close the two principles of ordinary induction and course-of-values induction are. However, consider the following informal re-statements of the two principles:

(1) For any property P that applies to 0, if a natural number has P whenever *its immediate predecessor* has P, then all natural numbers have P.

(2) For any property P, if a natural number has P whenever *all of its predecessors* have P, then all natural numbers have P.

Notice that the proviso in (1), that P apply to 0, is in effect built into (2) as well: the antecedent of (2) holds only if 0 has P. This is because 0 lacks predecessors in ω, and so it is always vacuously true that all predecessors of 0 have P.

Thus it is clear that (2) logically implies (1). Whether the converse holds is more difficult to see. Certainly the condition 'all of its predecessors have P' is, in general, much stronger than the condition 'its immediate predecessor has P'.

Hence the antecedent of (2) would appear to be weaker than that of (1) and, consequently, the whole principle (2) stronger than (1). We shall now prove that this appearance is deceptive: (1) logically implies (2), so the two principles are in fact equivalent. This is good to bear in mind, for it is often more convenient to use the course-of-values formulation than the ordinary one.

THEOREM 0.1.1. *The principle of course-of-values induction is correct.*

Proof. Take any $S \subseteq \omega$ and assume

(†) $\qquad \forall i(i < n \Rightarrow i \in S) \Rightarrow n \in S$, for all n.

For each n, let (A_n) be the statement $\forall i(i < n \Rightarrow i \in S)$. Thus the assumption (†) may be rewritten in the following form:

(††) $\qquad (A_n) \Rightarrow n \in S$, for all n.

If we could prove that, for every n, (A_n) is true, then it would follow from (††) that $S = \omega$, and so we would have proved the theorem. In other words, the proof of the theorem has been reduced to the proof of the claim that (A_n) is true for every n. This claim we now proceed to prove by ordinary induction.

Basic step. (A_0) is the statement $\forall i(i < 0 \Rightarrow i \in S)$. There is no natural number smaller than 0, so (A_0) is vacuously true.

Inductive step. Let n be any fixed number. Assume — as our inductive hypothesis — that

(§) $\qquad (A_n)$ is true.

But (§) is just another way of expressing the condition

$$\forall i(i < n \Rightarrow i \in S).$$

Furthermore, from (††) and (§) it follows that $n \in S$. Therefore,

$$\forall i(i < n + 1 \Rightarrow i \in S).$$

But this condition is nothing else but (A_{n+1}). Thus our induction is complete. ∎

0.2. Induction over other structures

A naïve but natural question is this: why does induction over the natural numbers work? Any answer, it seems, would have to deal with the structure of the set of natural numbers. This set can certainly be described simply enough: all one needs is the number 0 and the relation of immediate successor. Thus equipped one can describe any number one wishes by appealing to the immediate successor relation sufficiently many times:

1 is an immediate successor of 0 (in fact, the only one);
2 is an immediate successor of 1 (in fact, the only one);
3 is an immediate successor of 2 (in fact, the only one);

and so on.

It is not only in arithmetic that we may be confronted by an infinitude of objects for which we want to prove something. In this book we will often work with infinite sets of, for example, formulas. Also in such non-arithmetical cases we can often use induction by taking into account some way in which the objects in question can be described — or generated, as we shall say in this section.

By a relation we mean any set of tuples. (A tuple is simply an ordered set.) Usually a relation has a definite 'arity', meaning that all its tuples have the same number of elements. Thus a binary relation is a set of 2-tuples (ordered pairs), a ternary relation is a set of 3-tuples (ordered triples),..., a 17-ary relation is a set of 17-tuples, etc. But in the general case no assumption regarding arity is made. (A special case is that of zeroary relations. An empty ordered set would seem to be indistinguishable from an empty unordered set, and so there is only one empty tuple, viz., $\emptyset$, the empty set. Thus the one and only zeroary relation is $\{\emptyset\}$, which contains the empty tuple as its sole element. Strictly speaking, this relation is distinct from the empty relation, $\emptyset$, another degenerate case.)

To be quite specific we lay down the following definition: R is a *relation* (*in a set A*) if and only if, for any $z \in R$, either $z = \emptyset$ or

$$\exists n \geq 0 \exists a_0,\ldots, a_{n-1}, b \in A(z = \langle a_0,\ldots, a_{n-1}, b \rangle).$$

(Here we adopt the convention that if, in the latter case, $n = 0$ then $z = \langle b \rangle$.)

Let us say that $\langle B, R \rangle$ is an *inductive system* (*in* a set A) if

(i) $\qquad\qquad\qquad B \subseteq A$;
(ii) $\qquad\qquad\qquad$ R is a relation in A.

The elements of B we call the *basic objects* of the system, while R is called its *generating relation*. We define an infinite sequence $B_0, B_1,\ldots, B_k,\ldots$ of subsets of A in the following manner:

$$B_0 = B,$$

$$B_{k+1} = B_k \cup \{b : \exists n \exists a_0,\ldots, a_{n-1} \in B_k (\langle a_0,\ldots, a_{n-1}, b \rangle \in R)\}.$$

By the *set of objects generated by* $\langle B, R \rangle$ we mean the set

$$B_* = \bigcup_{k < \omega} B_k.$$

It should be noted that B_* is completely determined by $\langle B, R \rangle$ and, in particular, depends on R, even though there is nothing in the chosen symbolism to indicate the dependence on R. Also note that B_* need not be infinite. If B_* is finite, then there is some m such that, for all i, $B_m = B_{m+i}$.

The concept of inductive system generalizes the concept of natural numbers in the following sense. Let R_{nat} be the relation of immediate successor among natural numbers. Then $\langle \{0\}, R_{nat}\rangle$ is an inductive system which generates exactly the set of natural numbers: with respect to this system $\{0\}_* = \omega$. Now just as inductive systems may be regarded as a generalization of ω, so there is an induction principle concerning inductive systems that may be regarded as a generalization of ordinary induction.

Let us say that any rule R *preserves* a certain property P if whenever $\langle a_0,\ldots, a_{n-1}, b\rangle \in R$ and $a_0,\ldots, a_{n-1}$ have P, then also b has P. Suppose now that $\langle B, R\rangle$ is an inductive system, and that B_* is the set of objects generated by it. Then *the principle of induction over* $\langle B, R\rangle$ (or, less carefully, *over* B_*) may be expressed as follows. In words:

> If all basic objects have a certain property which the generating relation preserves, then all generated objects have that property.

In symbols:

$\forall b(b \in B \Rightarrow Pb)$ &

$\forall n \forall a_0,\ldots, a_{n-1}, b(\langle a_0,\ldots, a_{n-1}, b\rangle \in R \Rightarrow (Pa_0 \,\&\, \ldots \,\&\, Pa_{n-1} \Rightarrow Pb))$

$\Rightarrow$

$\forall b(b \in B_* \Rightarrow Pb).$

Let us say that a set S is *closed under* a relation R if whenever $\langle a_0,\ldots, a_{n-1}, b\rangle \in R$ and $a_0,\ldots, a_{n-1} \in S$, then $b \in S$. Using this terminology we can also give the following informal set-theoretical formulation of the principle of induction over B_*. In words:

> Any set that contains the basic objects and is closed under the generating relation also contains all generated objects.

In symbols:

$B \subseteq S$ &

$\forall n \forall a_0,\ldots, a_{n-1}, b(\langle a_0,\ldots, a_{n-1}, b\rangle \in R \Rightarrow (a_0,\ldots, a_{n-1} \in S \Rightarrow b \in S))$

$\Rightarrow$

$B_* \subseteq S.$

Thus we have actually one induction principle for each inductive system. By taking the whole class of inductive systems into account we could obtain a super-principle of induction:

> If $\langle B, R\rangle$ is an inductive system, then the principle of induction over $\langle B, R\rangle$ is correct.

It is clear that this super-principle logically implies the principle of ordinary induction since the set of natural numbers may be regarded as an inductive system. But also the converse implication holds: the super-principle can be reduced to induction of the ordinary kind. What this means is shown in the following theorem.

Let $\langle B, R \rangle$ be any inductive system, and let B_* be the set of objects generated by $\langle B, R \rangle$. Then, for every generated object a we define the *rank* of a (in $\langle B, R \rangle$) — written rank(a) — as follows:

$$\text{rank}(a) = 0, \quad \text{if } a \in B,$$
$$= k, \quad \text{if k is the smallest number } m > 0$$
$$\text{such that } a \in B_m - B_{m-1}.$$

Clearly, this defines a function $\quad \text{rank}: B_* \longrightarrow \omega$.

THEOREM 0.2.1. *The principle of induction over $\langle B, R \rangle$ is correct.*

Proof. Suppose that P is a property such that

(1)   Every basic object has P;
(2)   The generating relation preserves P.

We want to show that every generated object has P. This we will accomplish by ordinary induction on the rank of objects. What is meant by this phrase is this. For each n we introduce the statement

(R_n)   All elements of B_* of rank n or less have P.

We claim that (R_n) is true, for every n. If we can prove this claim, we shall have proved the theorem.

Basic step. The statement (R_0) is true, for the objects of rank 0 are the basic objects, and by assumption (1) they have P.

Inductive step. Let k be any fixed natural number. Assume, as our inductive hypothesis, that (R_k) holds:

(§)   All elements of B_* of rank k or less have P.

Let b be any object in B_* of rank $k + 1$. Then rank(b) is positive, and so $b \in B_{k+1} - B_k$. It follows that there are elements $a_0, \ldots, a_{n-1} \in B_k$, for some n, such that

(3) $\langle a_0, \ldots, a_{n-1} \rangle \in R$.

But every object in B_k has rank at most k. Hence $\text{rank}(a_0), \ldots, \text{rank}(a_{n-1}) \leq k$. By (§), then,

(4) $a_0, \ldots, a_{n-1}$ have P.

INDUCTION: SOME REMARKS 9

But by assumption (2), R preserves P. By (3) and (4), therefore, b has P. The induction is complete. ∎

The definition of B_* illustrates what has been called induction 'from the bottom up' (Enderton 1972). There is also induction 'from the top down' (Enderton 1972) which we must also consider. Let $\langle B, R \rangle$ be any inductive system in A. Then by the *set of objects induced by* $\langle B, R \rangle$ we mean the set

$$B^* = \bigcap \{S : B \subseteq S \ \& \ S \subseteq A \ \& \ S \text{ is closed under } R\}.$$

In other words, B^* is the smallest set that includes B and is closed under R.

This kind of definition is often expressed in the following slightly different way:

(1) Every basic object is an *induced object*.
(2) If $a_0, \ldots, a_{n-1}$ are *induced objects* and $\langle a_0, \ldots, a_{n-1}, b \rangle \in R$, then b is also an *induced object*.
(3) Nothing is an *induced object* except by virtue of (1) and (2).

Here each occurrence of the definiendum — 'induced object' — has been italicized. This is to emphasize that the definition consists of all three clauses: each contributes to the definition and is indispensable. Taken together they define B^*, the set of all induced objects (with respect to $\langle B, R \rangle$).

It is an important fact that induction from the bottom up and induction from the top down amount to the same thing: the set of generated objects coincides with the set of induced objects.

THEOREM 0.2.2. *Let $\langle B, R \rangle$ be an inductive system. Then $B_* = B^*$.*

Proof. The proof consists of two parts. First we prove that $B_* \supseteq B^*$. It follows from the definition of B_* both that B_* includes B and that B_* is closed under R. But by the definition of B^*, B^* is the smallest set to include B and be closed under R. Hence $B_* \supseteq B^*$.

Next we prove that $B_* \subseteq B^*$. Since $B_* = \bigcup \{B_k : k \in \omega\}$ it will be enough to prove that, for all k, $B_k \subseteq B^*$. This we do by ordinary induction.

Basic step. That $B_0 \subseteq B^*$ is clear, for $B_0 = B$ and B^* includes B.

Inductive step. Let k be any fixed number. As our inductive hypothesis, assume

(§) $$B_k \subseteq B^*.$$

Take any $b \in B_{k+1}$. Then there must be some elements $a_0, \ldots, a_{n-1} \in B_k$ such that $\langle a_0, \ldots, a_{n-1}, b \rangle \in R$. But, by (§), $a_0, \ldots, a_{n-1} \in B^*$. Furthermore, B^* is closed under R. Therefore $b \in B^*$. Consequently, $B_{k+1} \subseteq B^*$. The induction is complete. ∎

Presently we shall consider an application of the preceding. First, however, some definitions. If A is some given set, then we shall write

$$\text{id}_A = \{\langle x, x \rangle : x \in A\}$$

for what is often called the *diagonal* of A. Furthermore, by the *relative product* of two binary relations R and S in A we shall understand the binary relation

$$R|S = \{\langle x, y \rangle : \exists z (\langle x, z \rangle \in R \ \& \ \langle z, y \rangle \in S)\}.$$

This operation may of course be iterated any number of times. Thus we may define an infinite family of binary relations $R^0, R^1, \ldots, R^k, \ldots$ in A as follows:

$$R^0 = \text{id}_A$$
$$R^{k+1} = R^k | R.$$

R^k may be termed the *relative product of* R *by itself* k *times* or R *to the kth power*.

LEMMA 0.2.3. *Let R be a binary relation in A. Then, for every* k, $\langle x, y \rangle \in R^k$ *if and only if there are some elements* $z_0, \ldots, z_k \in A$ *such that* $z_0 = x$, $z_k = y$, *and, for all* $i < k$, $\langle z_i, z_{i+1} \rangle \in R$.

Proof. By induction on k. If $k = 0$, then the claim to be proved reduces to the claim that $\langle x, y \rangle \in \text{id}_A$ if and only if there are elements $z_0, z_1 \in A$ such that $z_0 = x$ and $z_1 = y$; that is, the trivial claim that $x = y$ if and only if $x = y$.

Next suppose that the claim holds for k. Then

$$\langle x, y \rangle \in R^{k+1}$$

iff $\exists u \in A(\langle x, u \rangle \in R^k \ \& \ \langle u, y \rangle \in R)$

iff $\exists u, z_0, \ldots, z_k \in A(z_0 = x \ \& \ z_k = u \ \&$

$\forall i < k (\langle z_i, z_{i+1} \rangle \in R) \ \& \ \langle u, y \rangle \in R)$

iff $\exists z_0, \ldots, z_k, z_{k+1} \in A(z_0 = x \ \& \ z_{k+1} = y \ \&$

$\forall i < k + 1 (\langle z_i, z_{i+1} \rangle \in R)).$ ∎

(Pedagogical note. The preceding proof is written in a style less careful but more common than the style we have been cultivating so far. The basic step and the inductive step are not announced by title. The induction hypothesis is indicated but not spelled out. In the inductive step it is not said exactly where this induction hypothesis is used. But by this time the reader should be able to supply any missing detail of this sort himself. The reader should also note our use of 'iff' as an abbreviation of 'if and only if' and of '∎' to mark the end of a proof.)

INDUCTION: SOME REMARKS

Now to the promised application. The concept of the ancestral of a binary relation is an important one, and there are two ways of defining it, both common. Let R be any binary relation in some given set A.

First definition. The *ancestral* of R is the relation

$$\underline{R} = \bigcup_{k<\omega} R^k.$$

Second definition. The *ancestral* of R is the reflexive transitive closure $\overline{R}$ of R. In other words, $\overline{R}$ is the smallest relation T such that

$$R \subseteq T,$$

$$\langle x, x \rangle \in T, \text{ for all } x \in A,$$

$$\langle x, y \rangle, \langle y, z \rangle \in T \Rightarrow \langle x, z \rangle \in T, \text{ for all } x, y, z \in A.$$

It would be possible to give a direct proof of the fact that these two definitions are equivalent; that is, that $\underline{R} = \overline{R}$. However, such a proof is already contained in the proof of Theorem 0.2.2; for the first definition is actually inductive from the bottom up, while the second definition is inductive from the top down. All this becomes evident if the situation is presented in the following terms. Let us introduce a new relation Q as follows:

$$Q = \{\langle\langle x, x \rangle\rangle : x \in A\} \cup \{\langle\langle x, y \rangle, \langle y, z \rangle, \langle x, z \rangle\rangle : x, y, z \in A\}.$$

Every element of Q is either a 1-tuple or a triple of ordered pairs of elements of A; and so, even though Q lacks a definite arity, it is nevertheless a relation in A x A. Moreover, as a binary relation, R is a subset of A x A. Therefore $\langle R, Q \rangle$ is an inductive system in A x A. Inspection shows that $\underline{R} = R_*$ and $\overline{R} = R^*$. But then, by Theorem 0.2.2, $\underline{R} = \overline{R}$.

In the sequel we will take it as obvious that the induction principles discussed in this chapter are correct, and we shall not always bother to distinguish between them. For example, the standard notation for the ancestral relation of a binary relation R will be R*, and a result like the following will be taken for granted:

LEMMA 0.2.4. *Suppose that* R *is a binary relation in some set* A, *and let* R* *be the ancestral of* R. *Then* $\langle x, y \rangle \in R^*$ *if and only if there are some elements* $z_0, \ldots, z_n \in A$ *such that* $z_0 = x$, $z_n = y$, *and, for all* $i < n$, $\langle z_i, z_{i+1} \rangle \in R$.

Proof. No matter which definition of ancestral is used, $R^* = \bigcup \{R^k : k < \omega\}$. Therefore the result follows from Lemma 0.2.3. ∎

0.3. Tree-structures and trees

It is easy to say what a *tree* is: a function whose domain is a tree-structure. This said, we shall concentrate on tree-structures and say no more about trees until much later in the book. (The notions of tree-structure and tree are needed only in Section 3.7. Although of independent interest, the material in the present section may therefore be omitted, without loss of continuity, by a reader who will omit also Section 3.7.) Tree-structures for our purposes are going to be certain sets of sequences of natural numbers. In dealing with these matters it will be convenient to use x, y, z as additional parameters for natural numbers, and **x, y, z** as parameters for sequences of natural numbers. Note that this goes beyond the general conventions declared elsewhere in the book.

A nonempty set T of nonempty finite sequences of natural numbers is a *tree-structure* if the following conditions are satisfied:

(i) There is a natural number z such that, whenever $\langle x_0,\ldots, x_m\rangle \in T$, then $x_0 = z$;
(ii) If $\langle x_0,\ldots, x_m, x_{m+1}\rangle \in T$, then $\langle x_0,\ldots, x_m\rangle \in T$;
(iii) If $\langle x_0,\ldots, x_m, k+1\rangle \in T$, then $\langle x_0,\ldots, x_m, k\rangle \in T$.

The following lemma is obvious, but it gives us an opportunity to study yet another kind of inductive argument.

LEMMA 0.3.1. *Let T be a tree-structure. Then* $\langle x_0,\ldots, x_m\rangle \in T$ *implies that, for all* $i \leq m$, $\langle x_0,\ldots, x_i\rangle \in T$. *Moreover,* $\langle x_0,\ldots, x_m, x_{m+1}\rangle \in T$ *implies that, for all* $k \leq x_{m+1}$, $\langle x_0,\ldots, x_m, k\rangle \in T$.

Proof. Assume that T is a tree-structure and that $\langle x_0,\ldots, x_m\rangle \in T$. We shall proceed by what is sometimes called *backward induction* (*from* m). In our terms this amounts to induction over the finite system $\langle B_m, R_m\rangle$, where $B_m = \{m\}$, and $R_m = \{\langle j+1, j\rangle : j < m\}$. The claim to be proved, then, is that, for all $i \leq m$,

(£) $\qquad\qquad\qquad \langle x_0,\ldots, x_i\rangle \in T.$

Basic step. For $i = m$, (£) reduces to the claim that $\langle x_0,\ldots, x_m\rangle \in T$, which is true by assumption.

Inductive step. Let j be any fixed natural number $< m$. Suppose, as the induction hypothesis, that (£) holds for $i = j+1$; that is, that

(§) $\qquad\qquad\qquad \langle x_0,\ldots, x_j, x_{j+1}\rangle \in T.$

We now wish to show that (£) holds also for $i = j$. But this is immediate, for, by condition (ii) in the definition of tree-structure, $\langle x_0,\ldots, x_j\rangle \in T$. The induction is complete.

The other part of the lemma is proved similarly. ∎

Thus, since T is nonempty, always $\langle z \rangle \in T$, where z is the element mentioned in condition (i) above.

Tree-structures are closely connected with inductive systems. If T is a tree-structure, there is in particular one inductive system, $\langle B_T, R_T \rangle$, of great interest, which we will call *the upward inductive system associated with* T. It is defined as follows:

$B_T = \{\langle z \rangle\}$, where z is the natural number referred to in (i);

$R_T = \{\langle\langle x_0,\ldots, x_m \rangle, \langle x_0,\ldots, x_m, x_{m+1} \rangle\rangle : \langle x_0,\ldots, x_m, x_{m+1} \rangle \in T\}$.

There is also another inductive system $\langle \overline{B}_T, \overline{R}_T \rangle$ which may be of interest under certain circumstances; for example, if T is finite. This system, which we call *the downward inductive system associated with* T, is defined as follows:

$\overline{B}_T = \{\langle x_0,\ldots, x_m \rangle \in T : \forall k(\langle x_0,\ldots, x_m, k \rangle \notin T)\}$;

$\overline{R}_T = \{\langle\langle x_0,\ldots, x_m, 0 \rangle,\ldots, \langle x_0,\ldots, x_m, q-1 \rangle, \langle x_0,\ldots, x_m \rangle\rangle :$
$\forall k(k < q \Leftrightarrow \langle x_0,\ldots, x_m, k \rangle \in T)\}$.

R_T and $\overline{R}_T$ are not inverses of one another (the inverse of a binary relation R is $\{\langle a, b \rangle : \langle b, a \rangle \in R\}$). However, a somewhat similar relationship exists and may be noted:

$\langle x_0,\ldots, x_{q-1}, y \rangle \in \overline{R}_T$ iff $\forall z(\langle y, z \rangle \in R_T$ iff $z \in \{x_0,\ldots, x_{q-1}\})$.

The next two theorems will show two senses in which induction over tree-structures is possible — always in the sense of Theorem 0.3.2, sometimes in the sense of Theorem 0.3.3.

THEOREM 0.3.2. *Let* T *be a tree-structure and* $\langle B_T, R_T \rangle$ *its upward associated system. Then* T *coincides with the set of objects induced by* $\langle B_T, R_T \rangle$.

Proof. Let T and $\langle B_T, R_T \rangle$ be as required. Let B* denote the set of objects induced by $\langle B_T, R_T \rangle$. We wish to show that T = B*.

That $T \supseteq B^*$ is clear — if a rigorous proof is called for, a straightforward induction over $\langle B_T, R_T \rangle$ suffices.

To see that $T \subseteq B^*$, it will be enough to prove that, for all m,

(£) if $\langle x_0,\ldots, x_m \rangle \in T$, then $\langle x_0,\ldots, x_m \rangle \in B^*$, for all $x_0,\ldots, x_m$.

This we do by induction on m.

Basic step. Suppose that $\langle x_0 \rangle \in T$. By condition (i) in the definition of tree-structure, $x_0 = z$, where z is the special natural number mentioned there and in the definition of B_T. Hence $\langle x_0 \rangle \in B_T$. Since $B_T \subseteq B^*$, this shows that (£) holds for m = 0.

Inductive step. Fix any j. As our induction hypothesis we assume:

(§)   If $\langle x_0, \ldots, x_j \rangle \in T$, then $\langle x_0, \ldots, x_j \rangle \in B^*$, for all $x_0, \ldots, x_j$.

Suppose now that

(1) $\langle y_0, \ldots, y_j, y_{j+1} \rangle \in T$.

By condition (ii) in the definition of tree-structure,

(2) $\langle y_0, \ldots, y_j \rangle \in T$.

Hence, by (§),

(3) $\langle y_0, \ldots, y_j \rangle \in B^*$.

By (1), (2), and the definition of R_T,

$$\langle y_0, \ldots, y_j \rangle R_T \langle y_0, \ldots, y_j, y_{j+1} \rangle.$$

By (3), then, $\langle y_0, \ldots, y_j, y_{j+1} \rangle \in B^*$. This shows that (£) holds for $m = j + 1$. The induction is complete, and so is the proof. ∎

By the *length* of a sequence we mean the number of elements in it; thus, the length of $\langle x_0, \ldots, x_m \rangle$ is $m + 1$.

THEOREM 0.3.3. *Let T be a tree-structure and $\langle \overline{B}_T, \overline{R}_T \rangle$ its downward associated system. Suppose that there is a finite upper bound to the length of the sequences of T. Then T coincides with the set of objects induced by $\langle \overline{B}_T, \overline{R}_T \rangle$.*

Proof. Let T and $\langle \overline{B}_T, \overline{R}_T \rangle$ be as required. In particular, suppose that $m + 1$ is the maximal length of any sequence in T; thus m is a natural number (possibly 0). Let B^* be the set of objects induced by $\langle \overline{B}_T, \overline{R}_T \rangle$. We wish to show that $T = B^*$.

As in the proof of the preceding theorem, it is clear that $T \supseteq B^*$. To see that $T \subseteq B^*$, it will be enough to prove that, for all $n \leq m$,

(£)   if $\langle x_0, \ldots, x_n \rangle \in T$, then $\langle x_0, \ldots, x_n \rangle \in B^*$, for all $x_0, \ldots, x_n$.

This we do by backward induction from m (cf. the proof of Lemma 0.3.1).

Basic step. If $\langle x_0, \ldots, x_m \rangle \in T$, then, by our assumption, for all k, $\langle x_0, \ldots, x_m, k \rangle \notin T$. Hence, by the definition of $\overline{B}_T$, $\langle x_0, \ldots, x_m \rangle \in \overline{B}_T$. This shows that (£) holds for $n = m$.

Inductive step. Fix any natural number $j < m$. As our induction hypothesis, assume that (£) holds for $n = j + 1$; that is, that

(§)   if $\langle x_0, \ldots, x_j, x_{j+1} \rangle \in T$, then $\langle x_0, \ldots, x_j, x_{j+1} \rangle \in B^*$, for all $x_0, \ldots, x_j, x_{j+1}$.

Take any $\langle y_0, \ldots, y_j \rangle \in T$. Let q be the smallest number such that, for all $k < q$, $\langle y_0, \ldots, y_j, k \rangle \in T$. If $q = 0$, then, for all k, $\langle y_0, \ldots, y_j, k \rangle \notin T$, and so, as in the basic step, $\langle y_0, \ldots, y_j \rangle \in \overline{B}_T$; *a fortiori*, $\langle y_0, \ldots, y_j \rangle \in B^*$. On the other hand, if

$q > 0$, then two things follow. First, by definition of $\overline{R}_T$,

$$\langle\langle y_0,\ldots, y_j, 0\rangle,\ldots, \langle y_0,\ldots, y_j, q-1\rangle, \langle y_0,\ldots, y_j\rangle\rangle \in \overline{R}_T.$$

Second, by (§), $\langle y_0,\ldots, y_j, k\rangle \in B^*$, for all $k < q$. Hence, $\langle y_0,\ldots, y_j\rangle \in B^*$. This shows that (£) holds for j, as we wanted. ∎

We now introduce some further terminology which will make it easier to reason about tree-structures. Suppose that T is a tree-structure, and let R_T be as defined above. The elements of T are often referred to as *nodes*. If x and y are nodes, then we say that x is an *immediate predecessor* of y, and that y is an *immediate successor* of x, if $x\ R_T\ y$ (hence x and y are distinct). Actually, if x is an immediate predecessor of y, then x is the unique immediate predecessor of y — a node may have many immediate successors, but never more than one immediate predecessor.

If $x\ R_T^k\ y$ (see Section 0.2 for the definition of relative product), then we say that x *precedes* y *by* k *steps* or that y *succeeds* x *by* k *steps*. If x precedes y by some finite but positive number of steps, then we say that x *precedes* y (x *is a predecessor of* y) or that y *succeeds* x (y *is a successor of* x). In this terminology a node does not precede or succeed itself, even though it both precedes and succeeds itself by zero steps. Notice that x is a predecessor of y, and hence y a successor of x, if and only if x and y are distinct nodes such that $x\ R_T^*\ y$, where R_T^* is the ancestral of R_T.

LEMMA 0.3.4. *Suppose that x and y are nodes of some tree-structure, and that* $x = \langle x_0,\ldots, x_m\rangle$. *If x precedes y by k steps, then there are natural numbers* $p_0,\ldots, p_{k-1}$ *such that* $y = \langle x_0,\ldots, x_m, p_0,\ldots, p_{k-1}\rangle$.

Proof. Let R_T be the generating relation of the upward inductive system associated with the tree-structure. Assume that $x = \langle x_0,\ldots, x_m\rangle$ and that $x\ R_T^k\ y$. We proceed by induction on k.

Basic step. If $k = 0$, then $x = y$, and no more need be said.

Inductive step. As our induction hypothesis, suppose that the lemma holds for $k = n$, where n is an arbitrary fixed natural number. Assume that $k = n + 1$. This means that $x\ R_T^{n+1}\ y$. Hence there exists a node z such that

(1) $\qquad\qquad\qquad x\ R_T^k\ z,$

(2) $\qquad\qquad\qquad z\ R_T\ y.$

By (1) and the induction hypothesis, $z = \langle x_0,\ldots, x_m, p_0,\ldots, p_{k-1}\rangle$, for some natural numbers $p_0,\ldots, p_{k-1}$. By (2), then, there is some natural number q such that $y = \langle x_0,\ldots, x_m, p_0,\ldots, p_{k-1}, q\rangle$. Hence the lemma holds for $k = n + 1$, and the induction is complete. ∎

Recall that we say of a binary relation R in an arbitrary set S that

R is *reflexive* (*in* S) iff $\forall a \in S(a\ R\ a)$,
R is *irreflexive* (*in* S) iff $\forall a \in S(\text{not } a\ R\ a)$,
R is *transitive* (*in* S) iff $\forall a, b, c \in S(a\ R\ b\ \&\ b\ R\ c \Rightarrow a\ R\ c)$,
R is *antisymmetric* (*in* S) iff $\forall a, b \in S(a\ R\ b\ \&\ b\ R\ a \Rightarrow a = b)$,
R is *weakly connected* (*in* S) iff $\forall a, b \in S(a \neq b \Rightarrow a\ R\ b$ or $b\ R\ a)$,
R is *connected* (*in* S) iff $\forall a, b(a\ R\ b$ or $b\ R\ a)$.

Furthermore, we say that

R is a *partial ordering* (*of* S) iff it is reflexive, transitive and antisymmetric in S,
R is a *linear ordering* (*of* S) iff it is a partial order of S and connected in S;
R is a *strict partial ordering* (*of* S) iff it is irreflexive and transitive in S,
R is a *strict linear ordering* (*of* S) iff it is a strict partial ordering of S and weakly connected in S.

In this terminology, R_T^* is always a partial ordering in T, but it is not in general a linear ordering. Similarly, the relation of being a predecessor, and the converse relation of being a successor, are always strict partial orderings but not in general strict linear orderings.

A node without successors is a *top*, a node without predecessors a *root*. Every tree-structure has exactly one root. All finite and some infinite tree-structures have tops. In the terminology laid down above, B_T is the (singleton) set of roots, while $\overline{B}_T$ is the set of tops.

A tree-structure is *finitary* if every node has at most finitely many immediate successors, *infinitary* if every node has infinitely many immediate successors. Obviously one and the same tree-structure can be both non-finitary and non-infinitary.

A subset $Q \subseteq T$ is said to be a *branch* (*through* the given tree-structure) if the following conditions are satisfied:

(i) If $x, y \in Q$, then $x\ R_T^*\ y$ or $y\ R_T^*\ x$;

(ii) If $\langle x_0, \ldots, x_m \rangle \in Q$ is not a top, then, for some k, $\langle x_0, \ldots, x_m, k \rangle \in Q$;

(iii) If $\langle x_0, \ldots, x_m, x_{m+1} \rangle \in Q$, then $\langle x_0, \ldots, x_m \rangle \in Q$.

In English, the import of these conditions is this: for any two distinct nodes of Q, one precedes the other; if Q contains a node, it also contains some immediate successor (unless it is a top); if Q contains a node, it also contains an immediate predecessor (unless it is the root). Thus, Q can be described as a subset of T which is linearly ordered by R_T^* (or, equivalently, strictly linearly ordered by the relation of being a predecessor) and which, furthermore, is maximal with respect to this property. We note, without proof, that every top determines a unique finite branch, and that, conversely, every finite branch determines a unique top.

LEMMA 0.3.5. *Let Q be a branch through any tree-structure. Then $\langle x_0, \ldots, x_m \rangle \in Q$ implies that, for all $i \leq m$, $\langle x_0, \ldots, x_i \rangle \in Q$.*

Proof. Analogous to that of Lemma 0.3.1. ∎

In the sequel we will consider sequences of various objects. In general the notations $\langle \alpha_i \rangle_{i<k}$ and $\langle \alpha_i \rangle_{i<\omega}$ will represent a finite sequence of k elements $\alpha_0, \ldots, \alpha_{k-1}$, respectively, an infinite sequence $\alpha_0, \ldots, \alpha_i, \ldots$. Sometimes, when it is desirable to leave open whether such a sequence is finite or infinite, it is convenient to employ the notation $\langle \alpha_i \rangle_i$. Similarly we might write $\{\alpha_i\}_i$ for the unordered set of elements of such a sequence, as we do, for example, in the following lemma.

LEMMA 0.3.6. *Let Q be a branch through any tree-structure. Then there is a sequence $\langle x_i \rangle_i$ of natural numbers such that $Q = \{\mathbf{x}_i\}_i$, where, for every i, we have $\mathbf{x}_i = \langle x_0, \ldots, x_i \rangle$.*

Proof. We define a sequence $\langle x_i \rangle_i$ of natural numbers as follows. Let i be any natural number. Define

$$x_i = y_i, \quad \text{if there is any node } y \in Q \text{ such that}$$
$$y = \langle y_0, \ldots, y_n \rangle \text{ and } n \geq i;$$
$$= \text{undefined}, \quad \text{otherwise.}$$

Contrary to appearance, this definition is correct. For let $y = \langle y_0, \ldots, y_n \rangle$ and $z = \langle z_0, \ldots, z_p \rangle$ be any two nodes of Q such that both $n, p \geq i$; we must now show that $y_i = z_i$. If $y = z$, this is certainly so; therefore assume that $y \neq z$. Since Q is a branch, it follows by the connectedness condition (i) in the definition of branch that one of y and z precedes the other. Without loss of generality, let us assume that y precedes z. By Lemma 0.3.4 then $n < p$, and $y_j = z_j$, for all $j \leq n$. Since $i \leq n$, this implies that $y_i = z_i$, as we wanted to show.

The sequence $\langle x_i \rangle_i$ having been defined, we define a corresponding sequence $\langle \mathbf{x}_i \rangle_i$ of nodes by stipulating that, for each i, $\mathbf{x}_i = \langle x_0, \ldots, x_i \rangle$. We claim that $Q = \{\mathbf{x}_i\}_i$.

To show that $Q \subseteq \{\mathbf{x}_i\}_i$, take any $y \in Q$. For some natural numbers $y_0, \ldots, y_n$, we have $y = \langle y_0, \ldots, y_n \rangle$. From the definition above it follows at once that $y_0 = x_0, \ldots, y_n = x_n$. Hence $y = \mathbf{x}_n$.

To prove that $Q \supseteq \{\mathbf{x}_i\}_i$, take any particular $\mathbf{x}_i$. Since x_i is defined, there must be some node $y = \langle y_0, \ldots, y_k \rangle$ in Q such that $k \geq i$. But the above definition then implies that $y_j = x_j$, for all $j \leq k$, so actually $y = \langle x_0, \ldots, x_k \rangle$. Since $y \in Q$, it follows from Lemma 0.3.5 that $\langle x_0, \ldots, x_i \rangle \in Q$. ∎

Having laid the groundwork for a theory of tree-structures, we will devote the remainder of this section to the proof of two theorems which will be of crucial importance in Section 3.7 below. The first is a classic result of long standing.

THEOREM 0.3.7. ('König's Lemma') *A finitary tree-structure is infinite only if it has an infinite branch.*

Proof. Assume that T is an infinite, finitary tree-structure. The nodes of T fall into two categories: those with at most finitely many successors, and those with infinitely many. Note the following two facts:

(1) The root of T has infinitely many successors;
(2) If a node has infinitely many successors, then also at least one of its immediate successors has infinitely many successors.

Assertion (1) follows from the assumption that T is infinite. To prove (2), let x be any node. Since T is finitary, x has at most finitely many *immediate* successors. If each of these has at most finitely many successors, then x, too, can have at most finitely many successors. This proves the contrapositive of (2).

We now define a sequence $\langle x_i \rangle_i$ of natural numbers as follows. First, let x_0 be the natural number guaranteed by condition (i) in the definition of tree-structure. Next, let j be any natural number and suppose that we have already defined $x_0, \ldots, x_j$. Then define x_{j+1} as the smallest number k such that $\langle x_0, \ldots, x_j, k \rangle$ is a node of T with infinitely many successors. By (1) and (2), this definition is correct. If we now define $\mathbf{x}_i = \langle x_0, \ldots, x_i \rangle$, for each i, it is immediate that $\langle \mathbf{x}_i \rangle_{i < \omega}$ is a branch, indeed an infinite one. ∎

Digression on König's Lemma. The preceding proof is of the kind that one occasionally encounters in classical mathematics or logic and which, although seemingly correct, may leave one with a feeling of uneasiness. The reason for this is probably that, in some sense, the proof fails to be 'constructive'. The crux is the distinction between nodes with at most finitely many successors and nodes with infinitely many. Classically the distinction makes sense; yet, how is one to identify the latter in a concrete case? It does not seem possible to give a general answer to this question, unless one can 'see' the whole infinite tree in one glance.

To elaborate this point, let us resort to simile. There is a certain infinite, finitary tree-structure in Olympos – a modern piece of sculpture called the Tree-Structure of Life – and one of the lesser gods is entrusted with the task of painting it: the nodes with at most finitely many successors are to be painted red, the nodes with infinitely many successors green. Now, being a lesser god, the painter has two handicaps. First, although he is able to move up and down the tree-structure at will, he can only move one step at a time, and slowly at that: there is a greatest lower bound to the time it takes to complete a step, whether he moves up (from node to immediate successor) or down (from node to immediate predecessor). Second, he is very nearsighted and not able to see much beyond the node at which he is. Thus, if he approaches a top node, he will not realize this till he gets to it: he can never know that a branch will not turn out to be finite, even if he has been climbing it for ever so long.

To some extent these limitations are offset by other assets — the slowness by immortality, the myopia by perfect memory. Even so it is clear that the painter is facing a task that exceeds his powers: the red-painting part of the job he can carry out, the green-painting part not. For every node with at most finitely many successors, the painter will eventually be able to identify it and paint it red. However, unless he has some independent information about the tree-structure, he will never need his green paint once he has painted the root. (There is one exception to this: if the painter has already painted a certain node green and all its immediate successors save one red, then he will infer that he can safely paint the remaining immediate successor green. Thus the only tree-structure in which the painter can perform well is one with a unique infinite branch. Note, however, that even this partial success is due to the information — which the painter is unable to check — that the tree-structure is infinite and finitary.)

The painter's problem depends on the 'subjective' facts about his own limitations as much as on any 'objective' fact about the tree-structure. It is instructive to compare our painter's situation with that of a major god, whose faculties presumably are infinite: such a god would immediately perceive which nodes have infinitely many successors.

This is not the place to do more than mention this interesting and important matter. Here the classical (or platonic) viewpoint is accepted without argument. According to this viewpoint, if a tree-structure is well-defined, then so is the set of successors of each node and, *ipso facto*, the cardinality of this set. Thus, as far as this book goes, epistemological qualms of the sort indicated are simply brushed aside. The painter has a problem all right, but it is a problem which is of no concern here.

At the heart of the epistemological problem besetting our proof of König's Lemma is the concept of the infinite. That concept is not involved in the following oft-quoted passage concerning the leaves of yesteryear. Nevertheless, the sentiment which it so forcefully airs is that of the classical viewpoint:

> So ist z. B. die Menge der Blüten, die ein gewisser, an einem bestimmten Orte stehender Baum im verflossenen Frühlinge getragen, eine angebliche Zahl, auch wenn sie niemand weiss; ein Satz also, der diese Zahl angibt, heisst mir eine objektive Wahrheit, auch wenn ihn niemand kennt.
> (Bernard Bolzano, *Wissenschaftslehre*, §25)

End of digression.

We turn now to the second of the two promised theorems, a result which bears a kind of duality relationship to the first: while König's Lemma states of every infinite, finitary tree-structure that some branch is infinite, the theorem to follow states of a certain infinitary tree-structure that every branch is finite.

First we introduce some technical terms. By a *virtually finite sequence* we shall understand a sequence $\langle x_i \rangle_{i<\omega}$ of natural numbers such that $\exists p \forall i > p(x_i = 0)$. Thus, from some point on, the sequence consists of 0's only. We shall represent such sequence in the following manner:

$$\langle x_0, \ldots, x_{p-1}, 0, 0, 0, \ldots \rangle.$$

If $x_{p-1} \neq 0$, then we say that p is the *(virtual) length* of the sequence and that x_p is the *mark*. By convention, the special sequence $\langle 0, 0, 0, \ldots \rangle$ is said to have length 0; it has no mark. Thus the mark, when it exists, is positive. Let us say of two virtually finite sequences $\mathbf{x} = \langle x_i \rangle_{i<\omega}$ and $\mathbf{y} = \langle y_i \rangle_{i<\omega}$ that $\mathbf{x}$ *dominates* $\mathbf{y}$ if $\exists q \forall i > q(x_q > y_q \ \& \ x_i = y_i)$. A *chain* of virtually finite sequences is simply a sequence $\langle \mathbf{x}_i \rangle_i$, finite or infinite, of virtually finite sequences. A *dominance chain* is a chain of virtually finite sequences such that, for every i, if both $\mathbf{x}_i$ and $\mathbf{x}_{i+1}$ are elements of the chain, then $\mathbf{x}_i$ dominates $\mathbf{x}_{i+1}$. Note that dominance is a transitive relation. Hence, if $\langle \mathbf{x}_i \rangle_i$ is a dominance chain, then $\mathbf{x}_i$ dominates $\mathbf{x}_{i+j}$, for all i and all positive j. If $\langle \mathbf{x}_i \rangle_i$ is a chain, then so is $\langle \mathbf{x}_{p+i} \rangle_i$, where p is a fixed natural number, and we will call it *the subchain generated from* $\langle \mathbf{x}_i \rangle_i$ *by* $\mathbf{x}_p$. Thus any chain is generated (from itself) by its first element. Note that a subchain generated from a dominance chain is itself a dominance chain.

THEOREM 0.3.8. *Any dominance chain of virtually finite sequences of natural numbers is finite.*

Proof. The proof will be carried out by a kind of double induction, with a primary and a secondary induction. The primary induction will be a course-of-values induction on the length of the generating sequence of a dominance chain. Let p be a fixed natural number and assume, as our first induction hypothesis:

(§)   Any dominance chain generated by a virtually finite sequence of length $<$ p is finite.

This reduces the proof of the lemma to the proof of this claim: any dominance chain generated by a sequence of length exactly p is finite.

We consider the cases $p = 0$ and $p > 0$ separately. If $p = 0$, then let C be any dominance chain of virtually finite sequences generated by a sequence of length 0. But the only sequence of length 0 is $\langle 0, 0, 0, \ldots \rangle$, and that sequence dominates no virtually finite sequence. Hence C consists of a single sequence and is very much a finite chain.

If $p > 0$, we face a more difficult situation; it is here that the secondary induction is required. More precisely, we will establish the remainder of our claim (that any dominance chain of virtually finite sequences is finite if it is generated by a sequence of length p) by course-of-values induction on the mark of the generating sequence.

Let q be any fixed positive number. As our second induction hypothesis we assume:

(§§)   Any dominance chain generated by a sequence of length p and mark $<$ q is finite.

Let $C = \langle \mathbf{x}_i \rangle_i$ be any dominance chain of virtually finite sequences $\mathbf{x}_i = \langle x_{i,j} \rangle_{j<\omega}$ such that $\mathbf{x}_0$ has length p and mark q. It will be enough to prove that C is finite.

There are two cases.

Case 1: There is a certain number $i > 0$ such that $x_{i,p-1} < q$. Let C' be the subchain of C that is generated by x_i. If $x_{i,p-1} = 0$, then the length of x_i is $< p$, and so, by (§), C' is finite. On the other hand, if $x_{i,p-1} > 0$, then the length of x_i is p, and its mark is $< q$. Hence, by (§ §), C' is finite. In either case, therefore, C' is finite. But the number of sequences that are in C but not in C' is i; so C, too, is finite.

Case 2: For every $i > 0$, we have $x_{i,p-1} \geqq q$. Since C is a dominance chain, and x_0 thus dominates every x_i with $i > 0$, this means that, for all i, $x_{i,p-1} = q$, and thus

$$x_i = \langle x_{i,0}, \ldots, x_{i,p-2}, q, 0, 0, 0, \ldots \rangle.$$

Now $q > 0$, so $q - 1 \geqq 0$. Hence we can define a new chain $\overline{C} = \langle \bar{x}_i \rangle_i$ by putting

$$\bar{x}_i = \langle x_{i,0}, \ldots, x_{i,p-2}, q-1, 0, 0, 0, \ldots \rangle.$$

It is easy to see that $\overline{C}$ is a dominance chain generated by the sequence

$$\bar{x}_0 = \langle x_{0,0}, \ldots, x_{0,p-2}, q-1, 0, 0, 0, \ldots \rangle.$$

If $q - 1 = 0$, then the length of $\bar{x}_0$ is $< p$, and so, by (§), $\overline{C}$ is finite. On the other hand, if $q - 1 > 0$, then the length of $\bar{x}_0$ is p, while the mark of $\bar{x}_0$ is $< q$. Hence, by (§ §), $\overline{C}$ is finite. But it is clear that C and $\overline{C}$ contain equally many elements (there is a one-to-one correspondence between the elements of C and the elements of $\overline{C}$); hence, C is finite. ∎

1
PROPOSITIONAL LANGUAGES

1.1. Propositional languages

The first thing to do is to agree on what a (propositional) language is. For our purposes a language might be identified with its propositional letters and (propositional) operators, because, as we shall see, they uniquely determine the formulas of the language. The operators are of two kinds, one that we call 'Boolean' and one that we call 'non-Boolean' or 'intensional'. But the specific nature (or 'shapes') of the letters and the operators need not concern us here as long as certain natural conditions are satisfied: for example, that there are operators and usually also letters, and that you can always recognize one when you see one; that they are all distinct, and that you can always recognize that they are; that every operator has a unique rank, and that you can always recognize what it is.

There is a strong tendency today to 'reduce' (as one says) as much as possible to sets or tuples of sets. In this vein we might define a *propositional language* as a quadruple $\mathcal{L} = \langle \text{Lett, Bop, Iop, r} \rangle$, where Lett (the set of *propositional letters*) and Bop (the set of *Boolean propositional operators*) and Iop (the set of *intensional propositional operators*) are sets, and r (the *rank function*) is a function from Bop $\cup$ Iop to the set ω of natural numbers $0, 1, 2, \ldots$ such that the following conditions obtain:

(i) Lett is denumerably infinite;
(ii) Bop $\cup$ Iop $\neq \emptyset$;
(iii) Lett $\cap$ Bop = Lett $\cap$ Iop = Bop $\cap$ Iop = $\emptyset$.

If $\oplus \in$ Bop $\cup$ Iop, then $r(\oplus)$ is the *rank* of $\oplus$. In general, an operator of rank n is said to be n-*ary*. The adjectives '1-ary', '2-ary', '3-ary',... are usually given Latinized readings: 'unary', 'binary', 'ternary',.... 0-ary ('zeroary') operators are called *propositional constants*. We shall often write Op for the set of all operators in $\mathcal{L}$; that is,

$$\text{Op} = \text{Bop} \cup \text{Iop}.$$

Thus (ii) implies that no language is so degenerate that it does not contain at least one operator.

The members of Lett and Op are the *primitive symbols* of $\mathcal{L}$. Sometimes one hears the term *alphabet* used for the set of primitive symbols.

It may be noted that the two cardinality assumptions (i) and (ii) differ somewhat in character. While assumption (ii) — that there is at least one operator — is made just to exclude what would otherwise be a trivial possibility, assumption (i) — that there are denumerably many propositional letters — really deprives us of a tiny piece of extra generality. For example, sometimes authors consider languages having but finitely many propositional letters; an extreme case is when there are none. Such cases are not directly covered here. However, a reader interested in languages violating (i) should be able to generalize our exposition easily enough.

1.2. Expressions

As has already been remarked, the set of natural numbers $0, 1, 2, \ldots$ is denoted by ω. We shall use lower case letters i, j, k, m, n, p, q as parameters ranging over ω. A *segment* of ω is a set $S \subseteq \omega$ with the following *convexity* property: if $i, i + 2 \in S$, then also $i + 1 \in S$. Segments of the form $I_n = \{i : i < n\}$ are called *initial* segments.

Let S be any non-empty set of objects. Then by a (finite) *sequence* of elements of S we mean a function from some initial segment I_n to S. Thus a sequence is a set of the form

$$\{\langle 0, x_0 \rangle, \ldots, \langle m-1, x_{m-1} \rangle\},$$

where $x_0, \ldots, x_{m-1} \in S$. We say that a sequence X *contains* an element x, or that x *occurs in* X, if there is some i such that $\langle i, x \rangle \in X$. One and the same element can of course occur more than once: $\{\langle 0, x \rangle, \langle 1, x \rangle\}$ is a perfectly good sequence. Since a set is completely determined by its elements, there is only one sequence containing no element at all, viz., the *empty sequence*, which we write $\emptyset$. Evidently the domain of $\emptyset$ is I_0.

New sequences can be formed from old ones by the binary operation of *concatenation*, which we write as $\star$. Thus if

(1) $\qquad X = \{\langle 0, x_0 \rangle, \ldots, \langle m-1, x_{m-1} \rangle\},$

(2) $\qquad Y = \{\langle 0, y_0 \rangle, \ldots, \langle n-1, y_{n-1} \rangle\}$

are sequences, then

(3) $X \star Y = \{\langle 0, x_0 \rangle, \ldots, \langle m-1, x_{m-1} \rangle, \langle m, y_0 \rangle, \ldots, \langle m+n-1, y_{m+n-1} \rangle\}$

is a uniquely defined sequence. It is readily seen that concatenation is an associative operation:

$$(X \star Y) \star Z = X \star (Y \star Z).$$

The set S^* of sequences of elements of S thus forms what is called a monoid (semigroup with identity) under concatenation. For, in addition to being closed on S^* and being associative, $\star$ has the empty sequence as an identity element:

$$X \star \emptyset = X = \emptyset \star X.$$

The validity of the following cancellation laws may be noted:

$$\text{If } X \star Z = Y \star Z, \text{ then } X = Y,$$

$$\text{If } X \star Y = X \star Z, \text{ then } Y = Z.$$

Note also that there is no nilpotent element:

$$X \star Y = \emptyset \text{ only if } X = Y = \emptyset.$$

In order to simplify the exposition we shall employ various conventions. Under these conventions, the sequences (1) and (2) would be written as

$$X = x_0 \star \ldots \star x_{m-1}, \text{ and } Y = y_0 \star \ldots \star y_{n-1}.$$

By the same token, the sequence under (3) would be written

$$X \star Y = x_0 \star \ldots \star x_{m-1} \star y_0 \star \ldots \star y_{n-1}.$$

We shall also allow hybrid forms of type

$$X \star z, \text{ and } z \star X,$$

where X is a sequence and z is an element. As $\star$ is associative, parentheses are not required for the notation to be meaningful; when, later, we sometimes use parentheses anyway, it is for heuristic purposes.

We may now define the concept of expression of a language simply enough: as a sequence of primitive symbols of the language. In other words, if $\mathcal{L} = \langle \text{Litt, Bop, Iop, r} \rangle$ is a language, then the set Exp of *expressions* of $\mathcal{L}$ is the set of sequences of elements of Litt $\cup$ Op. In what follows we shall use Roman upper case letters X, Y, Z, U, V as parameters ranging over Exp.

The *length* of an expression X, denoted $l(X)$, is the cardinality of the segment on which it is defined. Thus an expression is of length n if and only if it is defined on I_n. If X and Y are expressions of length m and n, respectively, then the length of $X \star Y$ is m + n. There are two special cases worth mentioning. One is expressions of length 1, which are of the form $\{\langle 0, x \rangle\}$. Often it is convenient to identify such a sequence with the sole element occurring in it. If we do — as we will here — then we can assert

$$\text{Litt, Op} \subseteq \text{Exp,}$$

which otherwise does not follow. The other special case we have already anticipated above: expressions of length 0, which are of the form $\{\ \}$. As we saw, there is only one empty expression, viz., $\emptyset$, the empty set.

We say that Y is a *subexpression of* X if there are U and V, not necessarily non-empty, such that

$$X = U \star Y \star V.$$

It is easy to see that the relation of being a subexpression is transitive. Note also that no account is taken of the 'position' in X where Y occurs: the *same* expression can occur at two *different* positions in the given expression X. If we wish to single out the position, we can use the integer $m = l(U)$ and say 'Y occurs beginning at position m in X.' Thus, if we wished to explain that two subexpressions Y and Y' *overlap as they occur in* X, where Y occurs at position m and Y' at position m', we would mean that

$$\text{either } m \leq m' < m + l(Y) \text{ or } m' \leq m < m' + l(Y).$$

Another approach would be to let the interval

$$\{i : m \leq i < m + l(Y)\}$$

represent the occurrence of Y in X. But usually we do not have to be quite as precise as this in giving rigorous definitions.

1.3. Formulas

We are now ready to define the set Form of *formulas* (of $\mathcal{L} = \langle \text{Litt, Bop, Iop, r} \rangle$). It is to be a certain subset of Exp, viz., the smallest set Σ that satisfies the following two conditions:

  (i) Every propositional letter is in Σ.
  (ii) If $A_0, \ldots, A_{n-1}$ are in Σ and $\oplus$ is an n-ary propositional operator, then the expression

$$\oplus \star A_0 \star \ldots \star A_{n-1}$$

is also in Σ.

(The same definition, expressed in the terms laid down in Chapter 0, would run as follows. Let R be the set of all ordered sets

$$\langle A_0, \ldots, A_{n-1}, \oplus \star A_0 \star \ldots \star A_{n-1} \rangle,$$

where $A_0, \ldots, A_{n-1}$ are expressions, and $\oplus$ is an n-ary operator. Then $\langle \text{Litt, R} \rangle$ is an inductive system in Exp, and Form is the set of generated (or induced) objects.)

The empty expression is not a formula, so formulas always have positive length. The formulas of length 1 are those that consist of either a propositional letter or a propositional constant. A formula of length 2 or more is always of the form $\oplus \star X$, where X is a non-empty expression and $\oplus$ is an n-ary operator, for some $n > 0$. We will use Roman upper case letters A, B, C, D as parameters ranging over the set of formulas.

The reader should note that our propositional language lacks special grouping devices: neither brackets nor parentheses nor commas are what we have called primitive symbols. They belong to the metalanguage in which our analysis is carried out, not to the object language (the formal propositional language) analysed.

This is a natural place to say something about notational conventions. After this chapter we shall usually prefer to write

$$\oplus[A_0,\ldots, A_{n-1}] \text{ for } \oplus \star A_0 \star \ldots \star A_{n-1}.$$

This is our official convention for n-ary operators, so to speak. Yet if $n \leq 2$ even the official convention is modified. Thus if $n = 0$, then we will usually write just

$$\oplus \quad \text{instead of} \quad \oplus[\ ].$$

If $n = 1$, then we will write

$$\oplus A \quad \text{instead of} \quad \oplus[A].$$

And if $n = 2$, then we will write

$$[A \oplus B] \quad \text{instead of} \quad \oplus[A, B].$$

Moreover, brackets will be handled in a casual manner: dropped when their absence does not invite confusion, then perhaps added again if need for their presence is felt. (This holds for parentheses, too, and not just in connection with formulas.) In other words, we shall conform to what is more or less standard practice.

We conclude this section by proving what is called a unique readability theorem: a theorem stating that a formula can be parsed in only one way. To do this we need the following lemma.

LEMMA 1.3.1. *Let A, B be formulas and X an expression. Then $A = B \star X$ only if $X = \emptyset$.*

Proof. Assume that $A = B \star X$. The proof is by induction on the length of A.

If $l(A) = 1$, then A consists of exactly one primitive symbol (a propositional letter or a propositional constant). But B, a formula, is non-empty. Hence $X = \emptyset$.

If $l(A) = n + 1$, for some $n > 0$, then suppose as the induction hypothesis that the lemma holds for all formulas of length at most n. That is to say, we assume the truth of the following claim:

(§)   if C, D are formulas and Y an expression, and if $l(C) \leq n$, then $C = D \star Y$ only if $Y = \emptyset$.

By an observation above, since $l(A) > 1$, the first symbol of A must be some operator $\oplus$. The rank of $\oplus$ is well defined; suppose that it is k. Since A is a formula there must be some formulas $C_0, \ldots, C_{k-1}$ such that

$$A = \oplus \star C_0 \star \ldots \star C_{k-1}.$$

But B has the same first symbol as A. Hence there are some formulas $D_0, \ldots, D_{k-1}$ such that

$$B = \oplus \star D_0 \star \ldots \star D_{k-1}.$$

Suppose first that $l(C_0) \geq l(D_0)$. Then there must be some expression such that $C_0 = D_0 \star Y$. But it is certain that $l(C_0) < l(A)$ and hence that $l(C_0) \leq n$. By the induction hypothesis (§), then, $Y = \emptyset$. Hence $C_0 = D_0$.

On the other hand, if $l(C_0) \leq l(D_0)$, then there is some expression Z such that $C_0 \star Z = D_0$. By the same argument as in the preceding paragraph, $Z = \emptyset$, and so $C_0 = D_0$. Hence in any case $C_0 = D_0$.

Repeating this argument $k - 1$ times, we conclude that $C_i = D_i$ for all $i < k$. Consequently, $A = B$ and $X = \emptyset$. ∎

THEOREM 1.3.2. *('Unique Readability'). Let $\oplus$ be a j-ary and $\triangle$ a k-ary operator. Suppose that*

$$X = \oplus \star A_0 \star \ldots \star A_{m-1}, \text{ and } Y = \triangle \star B_0 \star \ldots \star B_{n-1}$$

are expressions, where $A_0, \ldots, A_{m-1}, B_0, \ldots, B_{n-1}$ are formulas. Then $X = Y$ only if $\oplus = \triangle$, $m = n$, and, for all $i < m$, $A_i = B_i$. Moreover, X is a formula only if $j = m$.

Proof. The proof consists in applying Lemma 1.3.1 a suitable number of times. Note that it is immediately clear that $\oplus = \triangle$.

Either $l(A_0) \geq l(B_0)$ or $l(A_0) \leq l(B_0)$. In the former case there is some expression X' such that $A_0 = B_0 \star X'$. In the latter case there is some expression Y' such that $A_0 \star Y' = B_0$. In either case it follows from the lemma that $A_0 = B_0$. Repeating this argument $\min(m, n) - 1$ times, we conclude that $A_i = B_i$ for all $i < \min(m, n)$.

Now $m \neq n$ is impossible; since if, say, $m > n$ were the case, then because $X = Y$ and by cancellation of the $A_i = B_i$ for $i < n$, we would have

$$A_n \star \ldots \star A_{m-1} = \emptyset.$$

But as $A_n \neq \emptyset$, this is a contradiction.

Assume, finally that X is a formula. We may write, by definition,

$$X = \oplus \star C_0 \star \ldots \star C_{j-1},$$

for suitable choice of formulas C_i. But, by what we have just proved, $j = m$ must follow. ∎

1.4. Subformulas

The important notion of subformula is introduced in the following way. We say that B is an *immediate subformula of* A if $A = \oplus[C_0,\ldots, C_{n-1}]$, for some n-ary operator $\oplus$, and $B = C_i$, for some $i < n$. The relation 'subformula of' is then defined as the ancestral of 'immediate subformula of'; that is, as the reflexive, transitive closure of the latter relation. In other words, B is a *subformula of* A if and only if there are some formulas $D_0,\ldots, D_m$ such that the following conditions are satisfied:

(i) D_i is an immediate subformula of D_{i+1}, for all $i < m$;
(ii) $D_0 = B$;
(iii) $D_m = A$.

Thus, while A is not an immediate subformula of itself, yet it is a subformula of itself (the case when $m = 0$).

(If we wanted to insist on the terminology introduced in Chapter 0, then the same definition could be expressed as follows. Let R be the set of all ordered pairs

$$\langle \oplus \star C_0 \star \ldots \star C_{n-1}, C_i \rangle,$$

where $\oplus$ is an n-ary operator, $C_0,\ldots, C_{n-1}$ are formulas, and $i < n$. Then, for every A, $\langle \{A\}, R \rangle$ is an inductive system in Form, and the set of generated or induced objects – that is, the set $\{A\}_* = \{A\}^*$ – is the set of subformulas of A.)

We say that B *occurs in* A if B is a subformula of A.

A different and shorter definition of subformula than the one given would have been to stipulate that X is a subformula of A if X is a formula that is also a subexpression of A. Fortunately the two definitions are equivalent. But to prove that they are, we need some auxiliary results.

LEMMA 1.4.1. *Let* X, Y *be expressions and* $\oplus$ *an operator such that* $X \star \oplus \star Y$ *is a formula. Then there are expressions* Y_0, Y_1 *such that*

(i) $Y = Y_0 \star Y_1$;
(ii) $\oplus \star Y_0$ *is a formula.*

Proof. Let us write $A = X \star \oplus \star Y$. The proof is by induction on the complexity of the formula A. Because of $\oplus$, A cannot be just a propositional letter. Hence $A = \triangle \star B_0 \star \ldots \star B_{n-1}$, for some n-ary operator $\triangle$ and formulas $B_0,\ldots, B_{n-1}$. Assume, as the induction hypothesis, that the lemma holds for $B_0,\ldots, B_{n-1}$, the complexity of which is less than that of A.

First suppose that $X = \emptyset$. Then $\oplus = \triangle$ and $Y = B_0 \star \ldots \star B_{n-1}$. Taking $Y_0 = Y$ and $Y_1 = \emptyset$ makes the lemma true in this case.

Next suppose that $X \neq \emptyset$. In this case there is a unique $i < n$ as well as some expressions $\overline{X}, \overline{Y}$ such that the following is true:

(1) $\qquad B_i = \overline{X} \star \oplus \star \overline{Y};$

(2) $\qquad X = \triangle \star B_0 \star \ldots \star B_{i-1} \star \overline{X};$

(3) $\qquad Y = \overline{Y} \star B_{i+1} \star \ldots \star B_{n-1}.$

(As usual, our notation is to be understood in such a way that $X = \triangle \star \overline{X}$ if $i = 0$, while $Y = \overline{Y}$ if $i = n - 1$.) By the induction hypothesis, (1) implies that there are expressions $\overline{Y}_0, \overline{Y}_1$ such that $\oplus \star \overline{Y}_0$ is a formula and $\overline{Y} = \overline{Y}_0 \star \overline{Y}_1$. Thus putting $Y_0 = \overline{Y}_0$ and $Y_1 = \overline{Y}_1 \star B_{i+1} \star \ldots \star B_{n-1}$ makes the lemma true in this case. For, by (3), $Y_0 \star Y_1 = Y$. ∎

The preceding lemma enables us to prove a fact that may seem obvious but that needs proving: if two formulas that are subexpressions of the same expression overlap, then one is a subexpression of the other. The following is a precise formulation of this assertion:

LEMMA 1.4.2. *Let $A_0, \ldots, A_{k-1}$ and B be formulas. Suppose that B is a subexpression of $A_0 \star \ldots \star A_{k-1}$. Then B is a subexpression of A_i, for some $i < k$.*

Proof. Notice that it is not claimed that $A_0 \star \ldots \star A_{k-1}$ is formula (in fact, it never is if $k > 1$). Thus there is no loss of generality if we assume that B overlaps with A_0. We must prove that on this assumption B is a subexpression of A_0. If B is of length 1, this is obviously so. Assume therefore that B is of length more than 1. Then there is some operator $\oplus$ and some expression X such that

(1) $\qquad B = \oplus \star X.$

Since B overlaps with A_0 it is readily seen that $\oplus$ must occur in A_0. That is to say, there are U, V such that

(2) $\qquad A_0 = U \star \oplus \star V.$

Therefore, by Lemma 1.4.1, there are V_0, V_1 such that

(3) $\qquad V = V_0 \star V_1,$

(4) $\qquad \oplus \star V_0$ is a formula.

From (1) and (4) it follows by Lemma 1.3.1 that $B = \oplus \star V_0$. But then B is a subexpression of A_0 by (2) and (3). ∎

Now we are able to establish the promised equivalence of the two contemplated subformula definitions:

THEOREM 1.4.3. *Let A, B be formulas. Then B is a subformula of A if and only if B is a subexpression of A.*

Proof. **Only-if-part.** Assume that B is a subformula of A. By definition, an immediate subformula is a subexpression. But as the relation of being a subexpression is transitive, and as the relation of being a subformula is the transitive closure of the relation of being an immediate subformula, it follows that B is a subexpression of A. (The reader may wish to give an inductive proof using the formulas D_i mentioned at the beginning of this section.)

If-part. Assume that B is a subexpression of A. We proceed by induction on the complexity of A. Since A is a formula, we can write:

$$A = \oplus \star C_0 \star \ldots \star C_{k-1},$$

where $\oplus$ is a k-ary operator. If $k = 0$, then $A = B = \oplus$, and there is nothing to prove. If $k > 0$, and B occurs at position 0 in A, then we can argue by Lemma 1.3.1 that $A = B$. If B does not occur at the front of A, then B is a subexpression of $C_0 \star \ldots \star C_{k-1}$. Hence, by Lemma 1.4.2 B is a subexpression of some C_i for $i < k$. By the induction hypothesis, B is a subformula of C_i, which in turn is a subformula of A; therefore, B is also a subformula of A. ∎

We append one technical result which we shall want to refer to later. Like the preceding theorem it seems obvious enough. But in the present setting the lemma is non-trivial and must be proved.

LEMMA 1.4.4. *Let X, Y be expressions and A, B formulas. Then the expression* $X \star A \star Y$ *is a formula only if* $X \star B \star Y$ *is.*

Proof. By induction on the number $l(X) + l(Y)$.

Suppose first that $l(X) + l(Y) = 0$. Then $X = Y = \emptyset$, and so $X \star B \star Y = B$. But B is a formula all right.

Next let n be any fixed number. As the induction hypothesis we assume that

(§) for all U, V, if $l(U) + l(V) \leq n$, then $U \star A \star V$ is a formula only if $U \star B \star V$ is.

Suppose that $l(X) + l(Y) = n + 1$. Thus the formula $X \star A \star Y$ is of length at least 2. Hence there is some k-ary operator $\oplus$, for some $k > 0$, and some formulas $C_0, \ldots, C_{k-1}$ such that

$$X \star A \star Y = \oplus \star C_0 \star \ldots \star C_{k-1}.$$

If $X = \emptyset$, then by Lemma 1.3.1, also $Y = \emptyset$, contradicting the assumption that $l(X) + l(Y) > 0$. Hence $X \neq 0$. So there is some X_0 such that $X = \oplus \star X_0$ and

$$X_0 \star A \star Y = C_0 \star \ldots \star C_{k-1}.$$

Hence, by Lemma 1.4.2, there is some $i < k$ and expressions U, V such that $C_i = U \star A \star V$. Notice that then

$X = \oplus \star C_0 \star \ldots \star C_{i-1} \star U$, and $Y = V \star C_{i+1} \star \ldots \star C_{k-1}$.

But C_i is a formula, and $l(U) + l(V) < n$. Hence by the induction hypothesis (§), $U \star B \star V$ is a formula. So the situation is this: $\oplus$ is a k-ary operator, and $C_0, \ldots, C_{i-1}, U \star B \star V, C_{i+1}, \ldots, C_{k-1}$ are formulas. Hence the expression

$$\oplus \star C_0 \star \ldots \star C_{i-1} \star (U \star B \star V) \star C_{i+1} \star \ldots \star C_{k-1}$$

is a formula. But this formula may also be written

$$(\oplus \star C_0 \star \ldots \star C_{i-1} \star U) \star B \star (V \star C_{i+1} \star \ldots \star C_{k-1}).$$

So the formula in question is nothing else but $X \star B \star Y$. ∎

1.5. Substitution

Our final effort in this chapter will be to introduce the concept of substitution. Ordinarily this concept is explained to students in something like the following way: Let A be a given formula and P a given propositional letter. To substitute a formula B for P in A means to replace, in A, each occurrence of P by an occurrence of B.

So as to make this a little more precise, say that P has exactly k occurrences in A. What this means is that there exist $k + 1$ uniquely determined expressions $X_0, \ldots, X_k$, possibly empty but certainly not containing any occurrences of P, such that A can be written on the form

$$X_0 \star P \star X_1 \star P \star \ldots \star X_{k-1} \star P \star X_k.$$

By *the result of substitution in A of B for P*, denoted $A(P/B)$, we would then mean the formula

$$X_0 \star B \star X_1 \star B \star \ldots \star X_{k-1} \star B \star X_k;$$

for by repeated applications of Lemma 1.4.4 it follows that this expression is indeed a formula.

In a similar, although more complicated, fashion we could describe *the result of simultaneous substitution in A of B_0 for $P_0, \ldots, B_{n-1}$ for P_{n-1}*, denoted $A(P_0/B_0, \ldots, P_{n-1}/B_{n-1})$, where $P_0, \ldots, P_{n-1}$ are distinct propositional letters. Again Lemma 1.4.4 can be used to show that the expression in question is a formula.

It should be noticed that $A(P_0/B_0, \ldots, P_{n-1}/B_{n-1})$ need not be the same formula as $A(P_0/B_0) \ldots (P_{n-1}/B_{n-1})$. For example, if P, Q, R are distinct propositional letters, then

$$P(P/Q, Q/R) = Q, \text{ and } P(P/Q)(Q/R) = R.$$

A more uniform treatment of substitution is possible if we avail ourselves of the concept of substitution function. A *substitution function* (in Form) is a function s: Form $\to$ Form such that, for every n-ary operator $\oplus$ and all formulas

$A_0, \ldots, A_{n-1}$,

$$s(\oplus \star A_0 \star \ldots \star A_{n-1}) = \oplus \star sA_0 \star \ldots \star sA_{n-1}.$$

Or, using the more compact 'official' notation:

$$s(\oplus [A_0, \ldots, A_{n-1}]) = \oplus [sA_0, \ldots, sA_{n-1}].$$

If B is a formula, then we say that sB is a *substitution instance of* B.

Note that the set of substitution functions is closed under functional composition, ∘, and in fact is a monoid under ∘. For if s, t are substitution functions, then so is s ∘ t, the function defined by the condition that, for all A,

$$(s \circ t)A = t(sA);$$

furthermore, ∘ is associative: for all substitution functions s, t, u,

$$(s \circ t) \circ u = s \circ (t \circ u);$$

and, finally, for all substitution functions s,

$$s \circ e = s = e \circ s,$$

where e is the identity function, mapping every formula on itself.

Let us say that two substitution functions s, t *agree* on (the members of) a set Σ of formulas if and only if, for all $A \in \Sigma$, sA = tA. Thus s and t are the same substitution function — s = t — if and only if they agree on *all* formulas. Notice that every substitution is completely determined by its behaviour on the propositional letters. This is shown by the following result.

THEOREM 1.5.1. *Let s and t be any substitution functions, and suppose that they agree on some set Σ of propositional letters. Then s and t agree on any formula with propositional letters drawn exclusively from Σ.*

Proof. Let A be a formula with propositional letters in $\Sigma \subseteq$ Litt (that is, if P is a propositional letter occurring in A, then $P \in \Sigma$). We wish to show that sA = tA. This is done by induction on the complexity of A.

If A is itself a propositional letter in Σ, then s and t agree on A by assumption. Suppose therefore that $A = \oplus [B_0, \ldots, B_{n-1}]$, for some n-ary operator $\oplus$ and some $B_0, \ldots, B_{n-1}$. Furthermore, suppose that, for all $i < n$, $sB_i = tB_i$ (the induction hypothesis!). Then, using the defining characteristic of substitution functions twice and the induction hypothesis n times, we conclude that

$$sA = \oplus [sB_0, \ldots, sB_{n-1}]$$
$$= \oplus [tB_0, \ldots, tB_{n-1}]$$
$$= tA. \blacksquare$$

In effect we have now introduced the concept of substitution in two different ways. They are equivalent, though, and this is established in the following theorem. In particular the theorem guarantees that sA is the same formula as $A(P_0/sP_0,\ldots, P_{n-1}/sP_{n-1})$, provided that $P_0,\ldots, P_{n-1}$ are distinct propositional letters and s is a substitution function agreeing with the identity function on all other propositional letters.

THEOREM 1.5.2. *Let s be a substitution function and $P_0,\ldots, P_{n-1}$ distinct propositional letters. Suppose that $sQ = Q$ for all propositional letters $Q \notin \{P_0,\ldots, P_{n-1}\}$. Then, for all A,*

$$A(P_0/sP_0,\ldots, P_{n-1}/sP_{n-1}) = sA.$$

Proof. We shall only consider the case $n = 1$, as the general case is more difficult only from the notational point of view. Thus we shall be content to prove the following more special claim. Suppose that P is some fixed propositional letter and that s is a substitution function such that, for every propositional letter $Q \neq P$, $sQ = Q$. Then, we claim, it holds for all A that $A(P/sP) = sA$. The proof is by induction on A.

Suppose first that A is just a propositional letter, R say. If $R = P$, then $P(P/sP) = sP$, as claimed. If $R \neq P$, then $R(P/sP) = R = sR$, also as claimed.

Suppose next that $A = \oplus[B_0,\ldots, B_{n-1}]$, for some n-ary operator $\oplus$ and formulas $B_0,\ldots, B_{n-1}$. As induction hypothesis, assume that $B_i(P/sP) = sB_i$, for all $i < n$. It is clear that

$$A(P/sP) = (\oplus \star B_0 \star \ldots \star B_{n-1})(P/sP)$$
$$= \oplus \star B_0(P/sP) \star \ldots \star B_{n-1}(P/sP)$$
$$= \oplus[B_0(P/sP),\ldots, B_{n-1}(P/sP)].$$

On the other hand, it follows from the definition of substitution function that $sA = \oplus[sB_0,\ldots, sB_{n-1}]$. By the induction hypothesis, therefore, $A(P/sP) = sA$ in this case, too. ∎

One final piece of notation: If Σ is a set of formulas, then we shall write $s\Sigma = \{sA : A \in \Sigma\}$.

2
COMMON LOGICS

2.1. Logics

What kind of things are logics? No doubt the word 'logic' is used in many different senses, even in this book. In this section we shall first define one quite general sense of logic. We shall then go on to define a family of logics that are of special importance to us: 'common' logics.

Logicians divide formulas, as well as arguments, into valid and non-valid. Outside proof theory the preoccupation has been with formulas rather than arguments. Thus it is customary in some branches of propositional logic to identify a logic with a certain set of formulas, viz., those validated by the logic — the 'theorems' or 'theses'. But for others who look on the theses of a logic as something like the visible part of the iceberg, this is not enough: according to them it is by its set of valid arguments that a logic is known. Since a thesis can be thought of as a special kind of argument, the latter conception is more general than the former; therefore, we shall adopt the more general view here.

We shall begin on a high level of abstraction. Let Form be the set of formulas of some given propositional language. We shall define logics as sets, the elements of which are ordered pairs $\langle \Gamma, \Theta \rangle$ of sets of formulas. In other words, the condition that something, L, must satisfy in order to be a *logic* is

$$L \subseteq P(\text{Form}) \times P(\text{Form}).$$

Here, it should be noted, we depart somewhat from tradition. In our terminology, usually a pair $\langle \Gamma, \Theta \rangle$ would be regarded as an element of a logic only if Γ and Θ are finite sets (or sequences) of formulas (or even only if Γ is finite and Θ is either empty or a singleton — see, e.g., Kleene (1952), pp. 442 ff.).

If L is a logic and $\langle \Gamma, \Theta \rangle \in L$, then it will often be convenient to write this relationship in the more graphic form

$$\Gamma \vdash_L \Theta,$$

leaving out the subscript when clarity allows. If $\Gamma \vdash_L \Theta$ we say that Γ *entails* Θ (*in* L) or that Θ is *deducible from* Γ (*in* L). If both $\Gamma \vdash_L \Theta$ and $\Theta \vdash_L \Gamma$ we say that Γ and Θ are (*weakly*) *interdeducible* (in L), in symbols $\Gamma \dashv\vdash_L \Theta$. (The reader should take care, however, that $\dashv\vdash$ is not usually an equivalence relation when applied to sets of more than one element.)

It will be convenient to adopt a few conventions. Just as we use Roman upper case letters A, B, C, D to denote formulas, we will use Greek upper case letters $\Gamma, \Theta, \Delta, \Lambda, \Omega, \Sigma, \Xi$ to denote sets of formulas. Instead of

$$\{A_0,\ldots, A_{m-1}\} \vdash \{B_0,\ldots, B_{n-1}\}$$

we write

$$A_0,\ldots, A_{m-1} \vdash B_0,\ldots, B_{n-1}.$$

Similarly, instead of

$$\Gamma \cup \{A_0,\ldots, A_{m-1}\} \vdash \Theta \cup \{B_0,\ldots, B_{n-1}\}$$

we write

$$A_0,\ldots, A_{m-1}, \Gamma \vdash \Theta, B_0,\ldots, B_{n-1}.$$

For $\Gamma \vdash \emptyset$ and $\emptyset \vdash \Theta$ we may write $\Gamma \vdash$ and $\vdash \Theta$, respectively; thus $\vdash$ by itself is short for $\emptyset \vdash \emptyset$. For the complement (negative) of $\vdash$ we write $\nvdash$.

Let us say that a set Σ of formulas is

consistent (*in* L) if $\Sigma \nvdash$,

inconsistent (*in* L) if $\Sigma \vdash$,

implausible (*in* L) if $\nvdash \Sigma$,

plausible (*in* L) if $\vdash \Sigma$.

Furthermore, let us define the set Th(L) of *theses* (*of* L) as the set of formulas that are deducible from the empty set:

$$\text{Th}(L) = \{A : \vdash_L A\}.$$

Similarly, let us define the set Antith(L) of *antitheses* (*of* L) as the set of formulas that entail the empty set:

$$\text{Antith}(L) = \{A : A \vdash_L\}.$$

The concept of antithesis, although not unheard of, is not traditional. Later we shall see that in classical logic it is superfluous, as are the concepts plausible/implausible. But at this stage they are not definable in terms of their more familiar relatives, and they are needed for the sake of symmetry. Thus notice that A is a thesis if and only if $\{A\}$ is plausible, while A is an antithesis if and only if $\{A\}$ is inconsistent.

A terminological remark: the reader may have expected the term 'theorem' in place of or as a synonym to 'thesis'; the two are often used interchangeably. (Actually 'thesis' is a good deal less current than 'theorem'.) But in this book we take the view that theorems are formulas that can be proved in an *axiom system*.

So far we have not offered much in the way of explanation of this slightly bizarre concept of entailment. We all have some feeling for what it might mean that a formula (the 'conclusion') is deducible from a set (of 'premises'). But a set? Unfortunately, at this point — before we have introduced any semantics — there is not much explanation to offer. It may be helpful (but not very informative) to suggest that $\Gamma \vdash \Theta$ may be read 'if all of Γ, then at least some of Θ'. Perhaps it would be easier to begin with $\nvdash$, the complement of the relation $\vdash$, and read $\Gamma \nvdash \Theta$ as 'Γ avoids Θ'. Then $\Gamma \vdash \Theta$ could be read 'Γ doesn't avoid Θ'. In any case it must be emphasized that our technical notion of logic contains no strength whatever. Later we will consider special classes of logics, but at the present stage all possibilities are open. In particular — the readings notwithstanding — it is not excluded that there are formula sets Γ and Θ to make some or even all of the following statements true:

For all $A \in \Gamma$, $A \vdash \Theta$, yet $\Gamma \nvdash \Theta$;

For some $B \in \Theta$, $\Gamma \vdash B$, yet $\Gamma \nvdash \Theta$;

For all $A \in \Gamma$ there is some $B \in \Theta$ such that $A \vdash B$, yet $\Gamma \nvdash \Theta$.

On account of its misleading associations with the work of entailment logicians (see Anderson and Belnap 1975), the term 'entail' is not ideal for the notion described here. But there are few alternatives in ordinary English, and we want to reserve 'imply' for semantic notions. The term 'deducible' is also misleading, suggesting as it does that, where Θ is deducible from Γ, there exists some construction (a 'deduction') by virtue of which Θ is actually 'deduced' from Γ. But such suggestions are not intended.

2.2. Some common conditions on logics

So far, so good: a logic is a deducibility relation, a binary relation on the power set of Form. Next we wish to delineate more specific families of logics that are geared to our main interest, that of studying classical operators. Therefore we shall introduce a number of conditions, trying to bring out what is essential about classical deducibility. What we want eventually is an exact concept of classical logic that is both workable and in reasonable accord with the vast literature on the subject. But we will not actually reach this goal until the end of the book.

The first condition we consider is that of *reflexivity*:

(Refl) $\qquad \Sigma \vdash \Sigma,\quad$ provided that $\Sigma \neq \emptyset$.

This is a common condition. Indeed, with reservation for the fact that our formulation allows Σ to be infinite, reflexivity is one of the least controversial conditions. A logic satisfying it is said to be *reflexive*.

Related to reflexivity are the two equally non-controversial conditions *diagonality* and *overlap*:

(Diag) $A \vdash A,$ for every A.

(Overl) $\Gamma \vdash \Theta,$ provided that $\Gamma \cap \Theta \neq \emptyset$.

The following condition we call *monotonicity*: other traditional names are *thinning*, *dilution*, and *weakening*:

(Mono) If $\Gamma \vdash \Theta,$ then $\Gamma, \Gamma' \vdash \Theta, \Theta'$.

This condition may be viewed as the conjunction of two weaker conditions, *left-monotonicity*,

($Mono_L$) If $\Gamma \vdash \Theta,$ then $\Gamma, \Gamma' \vdash \Theta,$

and *right-monotonicity*,

($Mono_R$) If $\Gamma \vdash \Theta,$ then $\Gamma, \vdash \Theta, \Theta'$.

A logic satisfying the condition of monotonicity is called *monotonic*. Unlike reflexivity, monotonicity is a controversial condition. Among those who — for rather different reasons — find it unacceptable are relevance, quantum and inductive logicians. But classical logicians regard it as a correct principle.

We omit the proof of the following simple observation:

THEOREM 2.2.1. *In monotonic logics, the three conditions* (Refl), (Diag), *and* (Overl) *are equivalent.*

Next we consider some so-called *cut* conditions. One traditional looking cut condition is this:

(Cut_0) If both $\Gamma \vdash \Theta, C$ and $C, \Gamma' \vdash \Theta'$, then $\Gamma, \Gamma' \vdash \Theta, \Theta'$.

If we had adopted the convention that $\Gamma \vdash \Theta$ only if Γ and Θ are finite sets of formulas, then (Cut_0) would have been enough for our purposes. Indeed, that convention is adopted by almost everybody writing on this subject. But it is by no means necessary to do so, and we have not. Therefore, we need somewhat stronger cut conditions, viz., the following two:

(Cut_1) If both $\Gamma \vdash \Theta, \Omega$ and $A, \Gamma' \vdash \Theta'$, for every $A \in \Omega$, then $\Gamma, \Gamma' \vdash \Theta, \Theta'$.

(Cut_2) If both $\Gamma \vdash \Theta, B$, for every $B \in \Omega$, and $\Omega, \Gamma' \vdash \Theta'$, then $\Gamma, \Gamma' \vdash \Theta, \Theta'$.

Let (Cut) be the conjunction of (Cut_1) and (Cut_2). Both (Cut_1) and (Cut_2) are generalizations of (Cut_0), and hence so is (Cut). Implicit in our remarks above is the observation that (Cut_0) is a generalization in its turn of the following cut condition, which is the one usually encountered in the literature ('G' is mnemonic for Gentzen):

(Cut$_G$) If $\Gamma, \Gamma', \Theta, \Theta'$ are finite and both $\Gamma \vdash \Theta, C$ and $C, \Gamma' \vdash \Theta'$, then $\Gamma, \Gamma' \vdash \Theta, \Theta'$.

A logic satisfying (Cut) is said to be *closed under cut* and is also called a *cut logic*. A very special case of the cut conditions is *transitivity*:

(Trans) If $A \vdash B$ and $B \vdash C$, then $A \vdash C$.

We take this opportunity to prove the following result which will be needed later. The proof is very simple, but at this point it may be instructive.

THEOREM 2.2.2. *Let L be closed under monotonicity and cut. Suppose that* f: Form $\to$ Form *is a function such that, for all* A, $A \dashv \vdash fA$. *Then*

$$\Gamma \vdash \Theta \text{ iff } f\Gamma \vdash f\Theta.$$

Proof. Here, of course, $f\Gamma = \{fA : A \in \Gamma\}$, and $f\Theta = \{fB : B \in \Theta\}$. We will prove the only-if part. Assume that

(1) $\Gamma \vdash \Theta$.

For every A, $fA \vdash A$. Hence, by monotonicity,

(2) $f\Gamma \vdash A$, for each $A \in \Gamma$.

By cut, (1) and (2) yield

(3) $f\Gamma \vdash \Theta$.

Similarly, for every B, $B \vdash fB$. Hence, by monotonicity,

(4) $B \vdash f\Theta$, for each $B \in \Theta$.

By cut, (3) and (4) yield $f\Gamma \vdash f\Theta$, which is what we wanted.
The proof of the converse is similar. ∎

Next we have the condition of *substitutivity*:

(Subst) If $\Gamma \vdash \Theta$, then, for every substitution function s, $s\Gamma \vdash s\Theta$.

A logic satisfying (Subst) is said to be *substitutional* or *closed under substitution*. It is a familiar condition, and it is satisfied by almost all logics in the literature. But not all! Lemmon (1977), for example, does not impose it, and Åqvist (1973) has actually proposed a logic which is not closed under substitution.

The four main conditions we have seen so far — (Refl), (Mono), (Cut), (Subst) — are basic as far as classical logic goes. So basic, in fact, that we will set aside a special term for logics that satisfy them: 'common', for want of a better suggestion. In other words, a *common* logic is a reflexive, monotonic cut logic that is closed under substitution. The reader should note the great symmetry of these conditions: if $\vdash$ gives a common logic, then so does the converse relation $\dashv$.

2.3. Further conditions on logics

Among the conditions that we have decided not to include in the definition of common logic, perhaps the most plausible candidate is the condition of *consistency*:

(Cons) $\qquad L \neq P(Form) \times P(Form).$

There is only one logic not satisfying this condition, *the inconsistent logic*. In this logic anything goes: anything is deducible from anything. Such extreme permissiveness would seem to make the inconsistent logic rather uninteresting. For this reason it is not very often discussed in the literature (except, sometimes, inadvertently!). Yet it is not negligible; if not very interesting in itself, it occupies a very special position with respect to other logics. We shall hear more about this below. But there can be no doubt that consistency, in the sense of (Cons), is universally regarded as a highly desirable property of logics. The first two results are obvious.

THEOREM 2.3.1. *A monotonic logic is inconsistent if and only if* $\vdash$ *holds.*

THEOREM 2.3.2. *If L is a monotonic cut logic, then the following conditions are equivalent*:

(i) *L is inconsistent*;
(ii) $Th(L) \cap Antith(L) \neq \emptyset$;
(iii) $Th(L) = Antith(L) = Form.$

A much more unusual condition, which is nevertheless satisfied by practically every system in the literature, is that of *assertivity*:

(Ass) $\qquad$ For some Ξ and Π, we have $\Xi \vdash$ and $\vdash \Pi$.

As with monotonicity and cut, (Ass) is really the conjunction of two simpler conditions, viz., *left-assertivity*,

(Ass$_L$) $\qquad$ For some Ξ, we have $\Xi \vdash$,

and *right-assertivity*,

(Ass$_R$) $\qquad$ For some Π, we have $\vdash \Pi$.

Thus, a logic is left-assertive if and only if there exists an inconsistent set of formulas, right-assertive if and only if there exists a plausible one. The felicity of the chosen terminology will perhaps be questioned when it is realized that a monotonic logic is assertive — an attractive property, as it will turn out — if and only if there is a set which is at the same time inconsistent and plausible. The next result is also obvious.

THEOREM 2.3.3. *Let L be a monotonic logic. Then the following three conditions are equivalent*:

(i) *L is assertive*;
(ii) Form $\vdash$ *and* $\vdash$ Form;
(iii) *For some* Δ, *we have* $\Delta \vdash$ *and* $\vdash \Delta$.

The condition of assertivity was articulated by Steven T. Kuhn in 1976, perhaps for the first time. Kuhn, who used the term 'regular' rather than 'assertive', worked with the formulations given by the following theorem.

THEOREM 2.3.4. *Let L be a reflexive, monotonic cut logic. Then L is left-assertive if and only if*

(i) $\qquad$ *if* $\Gamma \vdash B$, *for all* B, *then* $\Gamma \vdash$

('*a set is inconsistent if it entails every formula*').
 Similarly, L is right-assertive if and only if

(ii) $\qquad$ *if* $A \vdash \Theta$, *for all* A, *then* $\vdash \Theta$

('*a set is plausible if it is entailed by every formula*').

Proof. We prove the first half of the theorem. First suppose that L is left-assertive. Then there is some Ξ such that

(1) $\qquad\qquad\qquad \Xi \vdash.$

Suppose that $\Gamma \vdash B$, for all B. Then, in particular,

(2) $\qquad\qquad\qquad \Gamma \vdash B,$ for all $B \in \Xi$.

Applying (Cut$_2$) to (1) and (2), we conclude that $\Gamma \vdash$.

Conversely, suppose that (i) holds. By reflexivity, $B \vdash B$, for every B. Hence, by monotonicity, Form $\vdash B$, for every B. By (i), then, Form $\vdash$. ∎

Finitariness is another important condition:

(Fin) $\qquad \Gamma \vdash \Theta$ only if there are finite sets Γ_0, Θ_0 with $\Gamma_0 \subseteq \Gamma, \Theta_0 \subseteq \Theta$, and $\Gamma_0 \vdash \Theta_0$.

This condition is sometimes called *compactness*, but usually the latter term is reserved for a semantic property.

THEOREM 2.3.5. *Let L be a monotonic, finitary logic. Then a set is consistent in L if and only if all its finite subsets are. Similarly, a set is implausible in L if and only if all its finite subsets are.*

THEOREM 2.3.6. *A finitary monotonic logic satisfies* (Cut$_G$) *if and only if it satisfies* (Cut).

Proof. It is of course only the only-if-direction that needs proof. Suppose that L is finitary and monotonic, and that L satisfies (Cut$_G$). Assume that $\Gamma, \Gamma', \Theta, \Theta', \Omega$ are such that

(1) $\quad\quad\quad\quad\quad \Gamma \vdash \Theta, B$, for all $B \in \Omega$;

(2) $\quad\quad\quad\quad\quad \Omega, \Gamma' \vdash \Theta'$.

We must show that $\Gamma, \Gamma' \vdash \Theta, \Theta'$. Since L is finitary, (2) implies that there are finite $\Omega^\dagger \subseteq \Omega, \Gamma^\dagger \subseteq \Gamma', \Theta^\dagger \subseteq \Theta'$ such that

(3) $\quad\quad\quad\quad\quad \Omega^\dagger, \Gamma^\dagger \vdash \Theta^\dagger$.

Let $\Omega^\dagger = \{B_0, \ldots, B_{n-1}\}$ (where thus n = 0 if and only if $\Omega^\dagger = \emptyset$). By (1) and finitariness, there are finite $\Gamma_0, \ldots, \Gamma_{n-1} \subseteq \Gamma$ and $\Theta_0, \ldots, \Theta_{n-1} \subseteq \Theta$ such that

(4) $\quad\quad\quad\quad\quad \Gamma_i \vdash \Theta_i, B_i$, for each i < n.

We contend that, for each i $\leq$ n, the statement

$$\{B_j : i \leq j < n\}, \Gamma_0, \ldots, \Gamma_{i-1}, \Gamma^\dagger \vdash \Theta^\dagger, \Theta_0, \ldots, \Theta_{i-1},$$

which we denote by S(i), is true. The contention is proved by induction. S(0) reduces to (3) and so is true. Suppose S(i) holds, for any i < n. Then (Cut$_G$) can be applied to (4) and S(i), yielding S(i + 1). This proves the contention. Hence also S(n) is true, that is to say,

$$\Gamma_0, \ldots, \Gamma_{n-1}, \Gamma^\dagger \vdash \Theta^\dagger, \Theta_0, \ldots, \Theta_{n-1}.$$

But then, by monotonicity, $\Gamma, \Gamma' \vdash \Theta, \Theta'$, as we wanted to show. ∎

A much more special condition on logics is that of congruence. Let us say that an n-ary operator $\oplus$ is *congruential* (in a given logic) if whenever $A_0 \dashv\vdash A_0', \ldots, A_{n-1} \dashv\vdash A_{n-1}'$ then $\oplus[A_0, \ldots, A_{n-1}] \dashv\vdash \oplus[A_0', \ldots, A_{n-1}']$. A logic is *congruential* if all its operators are.

One reason that congruence is such an important concept is that congruential logics possess an interesting technical characteristic: If A and A' are two formulas that differ only in that A contains an occurrence of a formula B in the same place as A' has an occurrence of a formula B', then A and A' are interdeducible if B and B' are. A precise statement is given in the following theorem.

THEOREM 2.3.7. (*'Replacement of interdeducible formulas in congruential logics.'*) *Let L be a congruential logic. Suppose that* A, A', B, B' *are formulas and* X, Y *expressions such that*

$$A = X \star B \star Y \text{ and } A' = X \star B' \star Y.$$

Then $B \dashv\vdash B'$ only if $A \dashv\vdash A'$. More generally, if s and t are two substitution functions, where for all propositional letters P we have

$$sP \dashv\vdash tP,$$

then, for all formulas C,

$$sC \dashv\vdash tC.$$

Proof. It is clear that the more general form includes the first statement of the replacement principle, because — by a suitable choice of a propositional letter Q occurring neither in A nor in A' — we can find s and t where

$$A = sC \text{ and } A' = tC,$$

and moreover

$$B = sQ \text{ and } B' = tQ,$$

with C being the formula $X \star Q \star Y$. Indeed, as only one replacement is involved, $sP = tP$ can be assumed to hold for all propositional letters $P \neq Q$.

The more general form is then proved by induction on the complexity of C under the stated assumption about s and t. In case C is of length 1, the conclusion is obvious. For C of greater length we can write

$$C = \oplus [D_0, \ldots, D_{n-1}].$$

Having assumed that the result holds for the shorter formulas D_i, the conclusion for C follows at once by the properties of substitution functions and the assumption that L is congruential. ∎

In words we may say that if A' is obtained from A by replacing any number of occurrences of formulas (as subformulas of A) by occurrences of interdeducible formulas, then A and A' are themselves interdeducible. And these replacements can be made over and over again, in effect by taking compositions of substitutions. We leave to the reader the exercise of generalizing the theorem to pairs of sets of formulas.

2.4. Syntactic equivalence

Intuitively we should like to say that two logics — formulated in perhaps very different object languages — are 'the same' or at least 'equivalent' if they are 'equally strong', if they 'come to the same thing'. According to one intuitive idea, this would be the case if in the first place the languages in which they are formulated are intertranslatable — if what can be expressed in one language can be expressed in the other one, too — and secondly whenever an argument in one logic is valid, then its counterpart in the other is also valid. Let us try to make this idea precise.

Suppose that L and L' are logics in the languages $\mathcal{L}$ and $\mathcal{L}'$, the formula sets of which are Form and Form', respectively. Then we say that L and L' are *syntactically equivalent with respect to f and g* if and only if f: Form $\to$ Form' and g: Form' $\to$ Form are functions such that the following conditions are satisfied:

(i)   For all A $\in$ Form, A $\dashv\vdash_L$ gfA;

(ii)   For all B $\in$ Form', B $\dashv\vdash_{L'}$ fgB;

(iii)   For all $\Gamma, \Theta \subseteq$ Form, $\Gamma \vdash_L \Theta$ iff f$\Gamma \vdash_{L'}$ fΘ;

(iv)   For all $\Delta, \Lambda \subseteq$ Form', $\Delta \vdash_{L'} \Lambda$ iff g$\Delta \vdash_L$ gΛ.

(The notation hΣ, where Σ is a set and h a function defined on Σ, stands of course for the set $\{hA : A \in \Sigma\}$.)

L and L' are said to be *syntactically equivalent* if there are functions f and g with respect to which they are syntactically equivalent.

THEOREM 2.4.1. *Syntactic equivalence is an equivalence relation in the class of reflexive cut logics.*

Proof. The proof is simple, but it is of some interest to see what assumptions are needed, and where.

Reflexivity. Any reflexive logic is syntactically equivalent to itself with respect to i and i, where i is the identity function on the set of formulas (conditions (i) and (ii) reduce to the condition that A $\dashv\vdash$ A, for all formulas A, while conditions (iii) and (iv) reduce to logical truths).

Symmetry. Whenever any logics L and L' are syntactically equivalent with respect to some functions f and g, then L' and L are syntactically equivalent with respect to g and f.

Transitivity. Assume that L, L', and L'' are cut logics in languages $\mathcal{L}, \mathcal{L}'$, and $\mathcal{L}''$, respectively, such that

(1) L and L' are syntactically equivalent with respect to f_1 and g_1,
(2) L' and L'' are syntactically equivalent with respect to f_2 and g_2.

We claim that L and L'' are syntactically equivalent with respect to $f_2 f_1$ and $g_1 g_2$. We must check that the four conditions above hold.

For condition (i) take any formula A of $\mathcal{L}$. Then, by (1) and condition (i),

(3)   A $\dashv\vdash_L g_1 f_1 A$.

But $f_1 A$ is a formula of $\mathcal{L}'$, so by (2) and condition (i), $f_1 A \dashv\vdash_{L'} g_2 f_2 f_1 A$. Hence, by (1) and condition (ii),

(4) $\quad\quad\quad\quad\quad\quad g_1f_1A \dashv\vdash_L g_1g_2f_2f_1A.$

By (3), (4), and cut, $A \dashv\vdash_L g_1g_2f_2f_1A$. Since functional composition is associative, this may be written $A \dashv\vdash_L (g_1g_2)(f_2f_1)A$, which is what we wanted.

For condition (iii), assume that Γ, Θ are sets of formulas of $\mathcal{L}$. Note that then $f_1\Gamma, f_1\Theta$ are sets of formulas of $\mathcal{L}'$ and $f_2f_1\Gamma, f_2f_1\Theta$ sets of formulas of $\mathcal{L}''$. Now, it follows from (1) and condition (iii) that $\Gamma \vdash_L \Theta$ iff $f_1\Gamma \vdash_{L'} f\Theta$, and from (2) and condition (iii) that $f_1\Gamma \vdash_{L'} f_1\Theta$ iff $f_2f_1\Gamma \vdash_{L''} f_2f_1\Theta$. Hence trivially $\Gamma \vdash_L \Theta$ iff $f_2f_1\Gamma \vdash_{L''} f_2f_1\Theta$.

Conditions (ii) and (iv) are verified in the same way as (i) and (iii), respectively. ∎

We remark that in case the logic is also monotonic the discussion can be simplified in view of Theorem 2.2.2. The simplification is that conditions (i)-(iv) above can be replaced by these two conditions:

(i') $\quad\quad\quad\quad\quad\quad \Gamma' \vdash f\Theta$ iff $g\Gamma' \vdash \Theta$;

(ii') $\quad\quad\quad\quad\quad\quad f\Gamma \vdash \Theta'$ iff $\Gamma \vdash g\Theta'$,

where $\Gamma, \Theta \subseteq$ Form and $\Gamma', \Theta' \subseteq$ Form'. The verification of this remark is left as an exercise.

The definition of syntactic equivalence relates to our intuitions as follows. The functions f and g are to be understood as translations of one language into the other. Conditions (i) and (ii) are a way of checking that the translations do their job — that at least they are, in a sense, inverses of one another. Conditions (iii) and (iv) are meant to guarantee that both translations preserve logical relationships. (Without semantics, our analysis is necessarily quite superficial. For fuller treatments, see, for example, Kanger (1968) and Kotas and Pietczkowski (1970).) In possession of this new concept we have yet another sense of the term 'logic'. Since syntactic equivalence is an equivalence relation, it partitions the class of all logics, in the sense of the previous section. A *logic in the wide sense* would simply be one of those equivalence classes.

2.5. Extensions

The word 'extension' may refer to either languages or logics, the latter use being the more frequent one. Suppose that $\mathcal{L} = \langle$Lett, Bop, Iop, r$\rangle$ and $\mathcal{L}' = \langle$Lett', Bop', Iop', r'$\rangle$ are languages such that

(i) $\quad\quad\quad\quad\quad\quad$ Lett $\subseteq$ Lett';
(ii) $\quad\quad\quad\quad\quad\quad$ Bop $\subseteq$ Bop';
(iii) $\quad\quad\quad\quad\quad\quad$ Iop $\subseteq$ Iop';
(iv) $\quad\quad\quad\quad\quad\quad$ r and r' agree on Bop $\cup$ Iop.

Then we say that $\mathcal{L}$ is a *sublanguage of* $\mathcal{L}'$ or that $\mathcal{L}'$ is an *extension of* $\mathcal{L}$. If L and L' are logics in $\mathcal{L}$ and $\mathcal{L}'$, respectively, and, in addition to (i)-(iv), also

(v) $\qquad L \subseteq L',$

then we say that L is a *sublogic of* L' or that L' is an *extension of* L. Furthermore, an extension L' of L is *conservative* (*over* L) if

(vi) $\qquad L = L' \cap P(\text{Form}) \times P(\text{Form}).$

An extension L' of L is *definitional* (*over* L) if

(vii) $\qquad \text{Lett} = \text{Lett}',$

(viii) for every n-ary operator $\oplus$ in $\mathcal{L}'$ but not in $\mathcal{L}$ and for all $P_0, \ldots, P_{n-1} \in$ Lett there is some $B_\oplus \in$ Form containing occurrences of $P_0, \ldots, P_{n-1}$ (and perhaps also other propositional letters) such that the following condition holds: whenever E and F are expressions and $C_0, \ldots, C_{n-1}$ are formulas such that $E \star \oplus [C_0, \ldots, C_{n-1}] \star F$ is a formula of $\mathcal{L}'$, then

($\oplus_{df}$) $\quad E \star \oplus [C_0, \ldots, C_{n-1}] \star F \dashv\vdash_{L'} E \star B_\oplus(P_0/C_0, \ldots, P_{n-1}/C_{n-1}) \star F.$

Let us introduce the following notation. We write $B(C_0, \ldots, C_{n-1})$ for $B(P_0/C_0, \ldots, P_{n-1}/C_{n-1})$. Moreover, we write $A \underset{df}{\sim} B$ if there are expressions E, F and formulas $C_0, \ldots, C_{n-1}$ such that

$$A = E \star \oplus [C_0, \ldots, C_{n-1}] \star F,$$
$$B = E \star B_\oplus(C_0, \ldots, C_{n-1}) \star F.$$

Thus what ($\oplus_{df}$) amounts to is the condition that, for all formulas A, B of $\mathcal{L}'$,

(RDE) $\qquad A \underset{df}{\sim} B$ implies $A \dashv\vdash_{L'} B;$

or, equivalently,

(RDE*) $\qquad A \underset{df}{\sim}^* B$ implies $A \dashv\vdash_{L'} B,$

where $\underset{df}{\sim}^*$ is the ancestral of $\underset{df}{\sim}$. We say that L' is closed under *replacement of definitional equivalents* if it satisfies these requirements.

THEOREM 2.5.1. *If L and L' are common logics such that L' is a conservative definitional extension of L, then L and L' are syntactically equivalent.*

Proof. Let i be the identity map from Form to Form'. We define g: Form' → Form as follows:

If P is a propositional letter, then define

$$g(P) = P.$$

If $\oplus$ is an n-ary operator in $\mathcal{L}$, then define, for all $A_0, \ldots, A_{n-1} \in$ Form',

$$g(\oplus[A_0, \ldots, A_{n-1}]) = \oplus[gA_0, \ldots, gA_{n-1}].$$

If $\oplus$ is an n-ary operator in $\mathcal{L}'$ but not in $\mathcal{L}$, then let B be a formula in Form of the kind guaranteed by the definition of definitional extension. Then define

$$g(\oplus[A_0,\ldots,A_{n-1}]) = B(P_0/gA_0,\ldots,P_{n-1}/gA_{n-1}).$$

It is a consequence of this definition and the fact that L' is a definitional extension of L that $A \underset{df}{\sim^*} gA$, for all $A \in \text{Form}'$ — an inductive argument proves this. Hence

(†) $\qquad\qquad A \dashv\vdash_{L'} gA, \quad \text{for all } A \in \text{Form}'.$

We must now check the four conditions in the definition of syntactical equivalence before we can conclude that L and L' are syntactically equivalent.

For condition (i), note that A and giA are interdeducible in L, for all $A \in \text{Form}$, since $giA = A$. For condition (ii), note that B and igB are interdeducible in L', for all $B \in \text{Form}'$, since $igB = gB$.

For condition (iii) we must prove that $\Gamma \vdash_L \Theta$ if and only if $i\Gamma \vdash_{L'} i\Theta$, for all $\Gamma, \Theta \subseteq \text{Form}$; that is, that $\Gamma \vdash_L \Theta$ if and only if $\Gamma \vdash_{L'} \Theta$. But this equivalence is immediate — in one direction because L' is an extension of L, in the other because the extension is conservative.

For condition (iv), suppose that, for any $\Delta, \Lambda \subseteq \text{Form}'$,

(1) $\qquad\qquad\qquad\qquad \Delta \vdash_{L'} \Lambda.$

By (†), $gA \vdash_{L'} A$ holds for all $A \in \text{Form}'$, hence in particular for all $A \in \Delta$. Therefore, by monotonicity,

(2) $\qquad\qquad\qquad g\Delta \vdash_{L'} A, \text{ for all } A \in \Delta.$

If cut is applied to (1) and (2), we obtain

(3) $\qquad\qquad\qquad\qquad g\Delta \vdash_{L'} \Lambda.$

By the same token, (†) yields $A \vdash_{L'} gA$, for all $A \in \text{Form}'$, hence in particular for all $A \in \Lambda$. Therefore, by monotonicity,

(4) $\qquad\qquad\qquad A \vdash_{L'} g\Lambda, \text{ for all } A \in \Lambda.$

Hence, if cut is applied to (3) and (4), we obtain $g\Delta \vdash_{L'} g\Lambda$. But L' is conservative over L; hence $g\Delta \vdash_L g\Lambda$. Thus $\Delta \vdash_{L'} \Lambda$ implies $g\Delta \vdash_L g\Lambda$. The converse implication is proved similarly. ∎

COROLLARY 2.5.2. *Two common logics are syntactically equivalent if there is a common logic that is a conservative definitional extension of both.*

Proof. A more careful formulation of the corollary would be this. Let L_1 and L_2 be common logics in languages $\mathcal{L}_1$ and $\mathcal{L}_2$, respectively. Let $\mathcal{L}^*$ be a language extending both $\mathcal{L}_1$ and $\mathcal{L}_2$. Furthermore, suppose that L^* is a common logic in $\mathcal{L}^*$ that is a conservative definitional extension of L_1 as well as of L_2. Then, it is claimed, L_1 and L_2 are syntactically equivalent.

Indeed, the claim follows easily. For, by Theorem 2.5.1, L_1 and L^* are syntactically equivalent, as are L_2 and L^*, and by Theorem 2.4.1, syntactic equivalence is an equivalence relation. ∎

2.6. The lemmas of Tukey and Lindenbaum

Readers of this book are likely to be familiar with the classic result that goes under the name of Lindenbaum's Lemma. This section is set aside for those who are not. Thus the entire section may well be skipped by knowledgeable readers.

We shall now speak of *properties of sets*, without saying exactly what a property is. A reader who feels uneasy about this may think of properties as collections of sets. Thus, our 'S has P' may be taken to mean '$S \in P$'; and so on. We say of a property P that it is *of finite character* if a set has P if and only if all of its finite subsets have. That is to say, P is of finite character if the following is always the case:

$$S \text{ has } P \text{ iff } \forall T \subseteq S \text{ (T is finite implies T has P)}.$$

Let X be any set and P a property of sets. We say that a subset of X is *maximal* (*for P over X*) if it has P and no strictly larger subset of X does. That is to say, a set $S \subseteq X$ is maximal (for P over X) if both the following conditions hold:

(i)   S has P;

(ii) $\forall T \subseteq X$ (T has P and $S \subseteq T$ imply $S = T$).

THEOREM 2.6.1. (*'Tukey's Lemma'*) *Let X be a set, and let P be a property of finite character. Then for every $S \subseteq X$ having P there is some $S^* \supseteq S$ that is maximal for P over X.*

Proof I. We give two proofs. The first one we carry out only for the case that X has denumerably many elements; a reader familiar with the concept of transfinite induction may try to extend the argument to the general case.

Assuming X to be denumerable, we know its elements can be enumerated. Let $a_0, a_1, \ldots, a_n, \ldots$ be such an enumeration. Hence every element of X will occur in this enumeration at least once.

The idea of this proof may be described as follows. We want to enlarge S into a maximal set having P. What could be more straightforward, then, than to add as many elements as possible, seeing to it that every element is considered for addition, yet taking care to preserve property P? To have control over the procedure, we will add only one element at the time; it is here the above enumeration is convenient, for it allows us to make sure every element is considered.

Setting out to implement this idea, we define an infinite family of sets S_n as follows:

$$S_0 = S;$$

$$S_{n+1} = S_n \cup \{a_n\}, \text{ if this set has P,}$$

$$= S_n, \text{ otherwise.}$$

It is easy to see, by induction on n, that each S_n has P: S_0 has it by assumption, and each S_{n+1} by construction.

Define

$$S^* = \bigcup_{n < \omega} S_n.$$

Obviously $S^* \supseteq S$. Notice that the S_ns are nested in the sense that for all m, $n \in \omega$, $S_m \subseteq S_n$ or $S_m \supseteq S_n$. We claim that S^* is the desired set; that is, maximal for P over X. So we must show two things: that S^* has P, and that no strictly larger subset of X does.

Suppose that T is any finite subset of S^*. In fact, suppose $T = \{t_0, \ldots, t_{k-1}\}$. For each $i < k$ take some number $f(i)$ such that $t_i \in S_{f(i)}$. Then $T \subseteq S_{f(0)} \cup \ldots \cup S_{f(k-1)}$. Let $m = \max\{f(0), \ldots, f(k-1)\}$. Then, since the S_ns are nested, it follows that $T \subseteq S_m$. But as we saw, S_m has P. Therefore, since T is finite and P is of finite character, also T has P.

The argument of the preceding paragraph shows that every finite subset of S^* has P. Hence, by the finite character of P, S^* has P.

Suppose now that T is a subset of X such that $T \supseteq S^*$ and T has P. We must show that $T = S^*$. To do this, take any $b \in T$. The given enumeration exhausts X, so b must occur somewhere in it; say $b = a_n$. Now $S_n \cup \{a_n\} \subseteq T$, and since T has P, so does $S_n \cup \{a_n\}$ (if a set has a property of finite character, then so does every subset!). But then $S_{n+1} = S_n \cup \{a_n\}$, and so $b = a_n \in S^*$. Therefore $T = S^*$.

Proof II. The following proof is a straightforward application of Zorn's Lemma and requires no assumption on the cardinality of X. A reader not familiar with this technique may skip this proof. (Help may be found in Halmos (1960) or Kelley (1955), two classic references.)

Let *P* be the collection of all subsets of X that contain S and have P. We note that *P* is non-empty, since $S \in P$. Note also that *P* is partially ordered by $\subseteq$, set inclusion.

Let *C* be a chain in *P*; that is, a subset of *P* such that for all T, $U \in C$, $T \subseteq U$ or $U \subseteq T$. Then $\bigcup C$ is an upper bound of *C*; that is, if $T \in C$ then $T \subseteq \bigcup C$ (for if $t \in T$, then trivially $t \in T \in C$). And $\bigcup C \in P$ (by an argument similar to one in proof I of this theorem). But what Zorn's Lemma states is that if every chain in a partially ordered, non-empty set has an upper bound in that set, then there is a maximal element in that set. In our case, therefore, there is some $S^* \in P$ such that, for all $T \in P$, if $S^* \subseteq T$ then $S^* = T$. Hence, S^* is maximal for P over X. ∎

Digression on Tukey's Lemma. Both proofs of the preceding theorems are what is called 'non-constructive', a fact that may warrant a short digression. The word 'classical' which is often used in this book — and even went into its title — is ambiguous. Not only do we propose to analyse logics that are classical in a certain technical sense yet to be defined, but also we carry out the analysis in the spirit of classical logic, where 'classical' means roughly the same as 'traditional'. Classical logic recognizes many methods or principles which are spurned by more constructively minded logicians, particularly the mathematical intuitionists (see, for example, Heyting 1956). Most famous among those principles are perhaps *reductio ad absurdum* and the Axiom of Choice. The former principle consists in accepting, as a proof of a statement ϕ (in the metatheory) any proof that we are led to absurdity if we assume that ϕ doesn't obtain. The Axiom of Choice asserts that for any non-empty set S of non-empty sets there exists a function c (called a *choice function (on* S)) such that, for all $x \in S$, $c(x) \in x$. The important thing to notice here is that it is only the bare existence of c that is asserted: we are not told what other properties it may have beyond being a choice function. In particular, the Axiom of Choice does not specify a procedure for computing the value of c for given arguments.

Reductio ad absurdum is a familiar strategy, and we have already used it, unabashedly, several times. The Axiom of Choice we used for the first time in the preceding proofs. This comes out particularly clearly in Proof II: Zorn's Lemma is equivalent to the Axiom of Choice (again, see Kelley 1955, or Halmos 1960), and it is used in Proof II to establish the existence of a set S* whose identity remains unknown. But also for the general case of Proof I is the Axiom of Choice or some equivalent principle needed (the construction procedure requires a well-ordering of X). In the special case of Proof I that we considered above we did not invoke the Axiom of Choice since there, by hypothesis, X is denumerable, which means that there exists some bijective function $f : \omega \to X$ (that is, f is both injective (one-to-one) and surjective (on to X)), which thus enumerates X. Still, the reader must not be fooled by the 'construction procedure' described in Proof I: it does not provide a general method to decide, of a given element a, whether or not $a \in S^*$, and so constructivists will not be impressed by our proof. **End of digression.**

We have presented Tukey's Lemma in full in order to emphasize how general the main idea in it really is; however, in logic one usually encounters it in a more specialized form. For example:

COROLLARY 2.6.2. *Let L be a finitary, monotonic logic. If Γ, Θ are sets of formulas such that $\Gamma \nvdash \Theta$, then there is a maximal set $\Gamma^* \supseteq \Gamma$ such that $\Gamma^* \nvdash \Theta$.*

Proof. As always, we are assuming some object language as given. Let P be the property that a set Σ of formulas has if $\Sigma, \Gamma \nvdash \Theta$. (A possible reading of this would be, 'Σ and Γ together avoid Θ'.) We claim that P is a property of finite character. First, if $\Sigma, \Gamma \nvdash \Theta$, then, by monotonicity, $\Sigma', \Gamma \nvdash \Theta$, for each finite $\Sigma' \subseteq \Sigma$. Conversely, assume that $\Sigma, \Gamma \vdash \Theta$. Then, by finitariness, there are some finite subsets $\Sigma_0 \subseteq \Sigma, \Gamma_0 \subseteq \Gamma, \Theta_0 \subseteq \Theta$ such that $\Sigma_0, \Gamma_0 \vdash \Theta_0$; whence, by monotonicity, $\Sigma_0, \Gamma \vdash \Theta$. Thus, P is of finite character.

This in turn means that Tukey's Lemma can be applied. It is clear that $\emptyset$, the empty set, has P. Hence there is some Σ^* that is maximal for P over the set of formulas. Putting $\Gamma^* = \Gamma \cup \Sigma^*$, it is readily seen that no larger set than Γ^* can avoid Θ. ∎

Similarly one proves:

COROLLARY 2.6.3. *Let L be a finitary, monotonic logic. If Γ, Θ are sets of formulas such that $\Gamma \nvdash \Theta$, then there is a maximal set $\Theta^* \supseteq \Theta$ such that $\Gamma \nvdash \Theta^*$.*

By a maximal L-consistent set is of course meant a set maximal for consistency over the set of formulas.

COROLLARY 2.6.4. *('Lindenbaum's Lemma'.) Let L be a finitary, monotonic logic. Every L-consistent set is contained in a maximal L-consistent set.*

Proof. Suppose Σ is consistent in L. Then $\Sigma \nvdash$. Hence, by Corollary 2.6.2, there is a maximal set $\Sigma^* \supseteq \Sigma$ such that $\Sigma^* \nvdash$. Since Σ^* is L-consistent, it follows that Σ^* is maximal also for L-consistency. (This result is also a direct corollary of Tukey's Lemma and Theorem 2.3.5.) ∎

We shall also need the following less common version of Lindenbaum's Lemma:

COROLLARY 2.6.5. *Let L be a finitary, monotonic logic closed under cut. Suppose that $\Gamma \nvdash \Theta$. Then there are sets $\Gamma^* \supseteq \Gamma$ and $\Theta^* \supseteq \Theta$ such that $\Gamma^* \cup \Theta^*$ contains all formulas, and yet $\Gamma^* \nvdash \Theta^*$. Hence, for each A, either A, $\Gamma^* \vdash \Theta^*$ or $\Gamma^* \vdash \Theta^*$, A, but not both.*

Proof. Suppose that $\Gamma \nvdash \Theta$. By Corollary 2.6.2 there is some maximal set $\Gamma^* \supseteq \Gamma$ such that $\Gamma^* \nvdash \Theta$. By Corollary 2.6.3, then, there is some maximal set $\Theta^* \supseteq \Theta$ such that $\Gamma^* \nvdash \Theta^*$. Suppose, by way of contradiction, that there is some $A \notin \Gamma^* \cup \Theta^*$. Then, by the maximality of Γ^*, we have A, $\Gamma^* \vdash \Theta$. Hence, by monotonicity,

(1) $\qquad\qquad\qquad A, \Gamma^* \vdash \Theta^*.$

By the maximality of Θ^*, we have

(2) $\qquad\qquad\qquad \Gamma^* \vdash \Theta^*, A.$

But then cut, applied to (1) and (2), yields $\Gamma^* \vdash \Theta^*$, which is absurd. ∎

3
BOOLEAN LOGICS

3.1. Boolean operators

The languages of classical propositional logic come with their operators divided into two mutually exclusive groups, those we call Boolean and those we call non-Boolean or intensional (with an 's'!). From now on the distinction will be important. Throughout the chapter we assume some given language $\mathcal{L} = \langle \text{Lett, Bop, Iop, r} \rangle$ with respect to which the discussion is carried out.

A formula is a *Boolean combination* (also: *Boolean compound*) *of formulas drawn from* the set $\{A_0, \ldots, A_{k-1}\}$ if it is a member of the closure under Boolean operators of the set $\{A_0, \ldots, A_{k-1}\}$; that is, the smallest set Σ such that

(1) $A_0, \ldots, A_{k-1} \in \Sigma$;
(2) If ☆ is any n-ary Boolean operator and $B_0, \ldots, B_{n-1} \in \Sigma$, then ☆$[B_0, \ldots, B_{n-1}] \in \Sigma$.

A formula B is a *Boolean combination* – or *compound* – *of* $A_0, \ldots, A_{n-1}$ if B is a Boolean combination of formulas drawn from $\{A_0, \ldots, A_{n-1}\}$ and, furthermore, $A_0, \ldots, A_{n-1}$ actually occur in B. We say that a formula is *purely Boolean* if it is a Boolean combination of propositional letters; that is, if it contains no non-Boolean operators.

By a *Boolean atom* we mean a formula that is either a propositional letter or of the form $\oplus[A_0, \ldots, A_{n-1}]$, where the operator $\oplus$ is not Boolean; in the latter case we speak of a *complex* Boolean atom (which only sounds like a *contradictio in adjecto*!). This terminology may be confusing in that there is nothing Boolean about Boolean atoms. 'Atom from the Boolean point of view' would have been a more informative term, even though we have not explained what the Boolean point of view is. What is important is that, from a certain standpoint, such a formula is something unanalysable, something indivisible: an atom.

The following simple fact is worth noting:

LEMMA 3.1.1. *Let A be any formula. Then there is a formula A_0 and a substitution function s such that*

(i) A_0 *is purely Boolean*;

(ii) *s is one-to-one*;
(iii) *sP is a Boolean atom, for every propositional letter* P;
(iv) *sP = P, for every propositional letter* P *in* A;
(v) $sA_0 = A$.

Proof. Any formula A may be written on a kind of normal form as follows: there is some number p, some expressions $X_0, \ldots, X_p$ and some complex Boolean atoms $C_0, \ldots, C_{p-1}$ such that, for all $i < p$, X_i contains no non-Boolean operator, and

$$A = X_0 \star C_0 \star \ldots \star C_{p-1} \star X_p.$$

Moreover, this representation is unique. (The existence claim may be proved by induction on A, the uniqueness claim with the help of Lemma 1.4.1. Notice that if A is a propositional letter, then $p = 0$ and $X_0 = A$. Moreover, if A is a complex Boolean atom, then $p = 1$, $X_0 = X_1 = \emptyset$ and $C_0 = A$.) Suppose that there are exactly m distinct formulas among $C_0, \ldots, C_{p-1}$, where thus $m \leq p$. Let $Q_0, \ldots, Q_{p-1}$ be a sequence of propositional letters not occurring in A such that, for all $i, j < p$, $C_i = C_j$ iff $Q_i = Q_j$. Thus the sequence $Q_0, \ldots, Q_{p-1}$ contains exactly m distinct propositional letters. We define A_0 and s as follows:

$$A_0 = X_0 \star Q_0 \star \ldots \star X_{p-1} \star Q_{p-1} \star X_p,$$

$$sP = C_i, \text{ if } P = Q_i, \text{ for some } i < p,$$

$$= P, \text{ otherwise}.$$

By Lemma 1.4.4, A_0 is a formula. It is clear that (i)–(v) are satisfied (for (v) invoke Theorem 1.5.2).

Note that A_0 is uniquely defined up to the choice of new propositional letters. ∎

To call an operator Boolean does not determine its behaviour. However, in this chapter we shall describe a class of logics which, roughly speaking, are those that treat Boolean operators in the manner they deserve. What is meant by this we shall explain presently, but only after we have introduced some rudimentary semantics.

3.2. Boolean semantics

In this book we assume that logic is two-valued. This assumption is by no means necessary, but it is made in classical logic. Thus we recognize two truth-values, the True and the False. For convenience we shall identify the True with the number 1 and the False with the number 0. This identification is of course arbitrary; any other representation would be admissible as long as the representatives of the True and the False are distinct.

BOOLEAN LOGICS

We let $\{0, 1\}^n$ be the Cartesian product of $\{0, 1\}$ by itself n times. That means that $\{0, 1\}^n$ is the set of all n-tuples of 0's and 1's. Note the two special cases when $n = 0$ and $n = 1$:

$$\{0, 1\}^0 = \{\langle \ \rangle\}, \quad \text{and} \quad \{0, 1\}^1 = \{\langle 0 \rangle, \langle 1 \rangle\},$$

where $\langle \ \rangle$ is the empty sequence and $\langle 0 \rangle$ and $\langle 1 \rangle$ are one-element sequences.

By an n-place *truth-value function* (in the literature usually called just 'truth function') we mean a function

$$\phi: \{0, 1\}^n \longrightarrow \{0, 1\}.$$

Thus an n-place truth-value function is a function assigning a truth-value to each n-tuple of truth-values. Hence, there are two 0-place truth-value functions, four 1-place truth-value functions,... – in general, there are 2^{n+1} n-place truth-value functions.

By a *(Boolean) matrix* (for a given language) we mean a function **M** defined on the set of Boolean operators, such that for each n-ary Boolean operator ☆, **M**(☆) is an n-ary truth-value function. By a *truth-value assignment for* a set Σ of formulas we simply mean a function from Σ to $\{0, 1\}$.

Suppose that **M** is a matrix and f is a truth-value assignment for some set Σ of formulas. Let Σ^* be the closure of Σ under all Boolean operators; that is, Σ^* is the smallest set to satisfy the two conditions

(1) $\Sigma \subseteq \Sigma^*$;
(2) For every n-ary Boolean operator ☆, if $A_0, \ldots, A_{n-1} \in \Sigma^*$, then also ☆$[A_0, \ldots, A_{n-1}] \in \Sigma^*$;

and thus Σ^* is the set of all Boolean combinations of formulas in Σ. The following then defines a particular truth-value assignment for Σ^*, the *extension* $f^\mathbf{M}$ *of* f *to* Σ^*, which agrees with f on Σ:

(i) For every $A \in \Sigma$, $f^\mathbf{M}(A) = f(A)$;
(ii) For every n-ary Boolean operator ☆,

$$f^\mathbf{M}(\text{☆}[A_0, \ldots, A_{n-1}]) = \mathbf{M}(\text{☆})(f^\mathbf{M}(A_0), \ldots, f^\mathbf{M}(A_{n-1})).$$

Note that if f is defined for every Boolean atom, then $f^\mathbf{M}$ is defined for every formula.

We say that a set Γ of formulas **M**-*implies* a set Θ of formulas, in symbols $\Gamma \models^\mathbf{M} \Theta$, if, for every truth-value assignment f to the set of Boolean atoms,

if $f^\mathbf{M}(A) = 1$ for every $A \in \Gamma$,

then $f^\mathbf{M}(B) = 1$, for some $B \in \Theta$.

We say that Γ and Θ are **M**-*equivalent*, in symbols $\Gamma \dashv\vdash^\mathbf{M} \Theta$, if Γ and Θ **M**-imply one another. We say, moreover, that a formula A is an **M**-*tautology*, if $\emptyset \models^\mathbf{M} A$; an **M**-*contradiction*, if $A \models^\mathbf{M} \emptyset$.

The implication concepts just defined exhibit a rather striking similarity with those of entailment. Thus the reader will understand at once what is meant by saying that we shall adopt for the implication concepts the same notational conventions as were laid down in section 2.1. But the similarity goes deeper, as the following result shows:

THEOREM 3.2.1. *Any matrix defines a consistent common logic.*

Proof. What is asserted is that if M is any fixed matrix, then the set $L = \{\langle \Gamma, \Theta \rangle : \Gamma \models^M \Theta\}$ is a logic that satisfies the five conditions (Cons), (Refl), (Mono), (Cut), and (Subst). We shall treat three of the conditions, (Cons), (Cut), and (Subst), leaving the remainder of the proof to the reader.

Consistency. By the definition of propositional language (see section 1.1), there is some n-ary operator $\oplus$ and some formulas $B_0, \ldots, B_{n-1}$ such that $A = \oplus[B_0, \ldots, B_{n-1}]$ is a formula. (If $n = 0$, then $\oplus$ is a constant and $A = \oplus$.) If $\oplus$ is non-Boolean, then it is easy to see that both $A \not\models^M$ and $\not\models^M A$, and so L is certainly consistent in this case. If $\oplus$ is Boolean, then, for any assignment f of truth-values to the Boolean atoms, either $f^M(A) = 1$ of $f^M(A) = 0$. In the former case $A \not\models^M$, in the latter case $\not\models^M A$. So in either case, L is consistent.

Cut. Suppose that $\Gamma, \Gamma', \Theta, \Theta', \Omega$ are such that

(1) $\qquad\qquad\qquad \Gamma \models^M \Theta, \Omega,$

(2) $\qquad\qquad\qquad C, \Gamma' \models^M \Theta',$ for every $C \in \Omega.$

We should like to prove that

(3) $\qquad\qquad\qquad \Gamma, \Gamma' \models^M \Theta, \Theta'.$

Let f be any assignment of truth-values for the set of Boolean atoms and assume, for *reductio ad absurdum*, that

(4) $\qquad\qquad\qquad f^M(A) = 1,$ for all $A \in \Gamma \cup \Gamma',$

(5) $\qquad\qquad\qquad f^M(B) = 0,$ for all $B \in \Theta \cup \Theta'.$

From (1) together with (4) and (5), it follows that there is some formula $C_0 \in \Omega$ such that $f^M(C_0) = 1$. But this, together with (4) and (5), shows that $C_0, \Gamma' \not\models^M \Theta'$, which contradicts the assumption (2). This argument shows that L satisfies (Cut_1). That L also satisfies (Cut_2) can be shown by a similar argument.

Substitution. Also this part we prove by a *reductio* argument: suppose

(1) $\qquad\qquad\qquad \Gamma \models^M \Theta,$

and yet

(2) $\qquad\qquad\qquad s\Gamma \not\models^M s\Theta,$

where s is some substitution function. By (2), there exists some truth-value assignment f for the set of Boolean atoms such that

(3) $\quad f^M(A) = 1$, for all $A \in s\Gamma$,

(4) $\quad f^M(B) = 0$, for all $B \in s\Theta$.

The following defines a new assignment of truth-values to Boolean atoms C: $g(C) = f^M(sC)$. We claim that

(5) $\quad g^M(C) = f^M(sC)$, for all $C \in$ Form.

This is easily proved by induction on C. If C is a Boolean atom, the claim is obviously true. If C is not a Boolean atom, then $C = \star[D_0, \ldots, D_{n-1}]$, for some n-ary Boolean operator $\star$ and formulas $D_0, \ldots, D_{n-1}$. Assuming, as our induction hypothesis, that (5) holds for $D_0, \ldots, D_{n-1}$, we get the following argument:

$$\begin{aligned} g^M(\star[D_0, \ldots, D_{n-1}]) &= M(\star)(g^M(D_0), \ldots, g^M(D_{n-1})) \\ &= M(\star)(f^M(sD_0), \ldots, f^M(sD_{n-1})) \\ &= f^M(\star[sD_0, \ldots, sD_{n-1}]) \\ &= f^M(s\star[D_0, \ldots, D_{n-1}]). \end{aligned}$$

Thus (5) holds for C. But (3) and (4) can be rewritten as

$$f^M(sA) = 1, \text{ for all } A \in \Gamma, \text{ and } f^M(sB) = 0, \text{ for all } B \in \Theta.$$

Hence, $g^M(A) = 1$, for all $A \in \Gamma$, and $g^M(B) = 0$, for all $B \in \Theta$. But this contradicts (1). ∎

We list the following fact for future reference.

COROLLARY 3.2.2. *Let M be any matrix. Suppose that s is a substitution function such that s is one-to-one and, for every propositional letter P, sP is a Boolean atom. Then, for all A, B, we have $A \vDash^M B$ iff $sA \vDash^M sB$.*

Proof. From left to right, the corollary follows from the preceding theorem. The converse is trivial. ∎

We said above that what we wish to do in this chapter is to single out a class of logics that treat Boolean operators 'in the manner they deserve'. Now it is time to return to that project. Let us say that a Boolean operator $\star$ *directly expresses* a truth-value function ϕ *under* a matrix M if $M(\star) = \phi$. In such a case we also say that ϕ is *directly expressible* under M. (Our conventions on matrices guarantee that ϕ is n-place where $\star$ is n-ary.) Now a logic would seem to treat a Boolean operator 'in the manner it deserves' if that operator 'expresses a truth-value function in L'. The question really becomes: how, using our newly gained semantic insights, can we make good syntactic sense out of the last vague suggestion? In particular, how can we eliminate the dependence on M?

The way out that we suggest here is as follows. Theorem 3.2.1 shows that there is a natural sense in which a matrix may be said to determine a logic. As will be seen, there is an important class of logics for which a kind of converse holds: each in a sense determines a unique matrix. For a truth-value function to be expressed in such a logic may reasonably be identified with being expressed under the corresponding matrix.

To make this precise we adopt the following terminology. If L is a logic and **M** is a matrix for the language of L, then we say that **M** *respects* L if it is never the case that $\Gamma \vDash^M \Theta$ while $\Gamma \nvdash_L \Theta$. In other words, **M** respects L if and only if, for all Γ and Θ, $\Gamma \vDash^M \Theta$ only if $\Gamma \vdash_L \Theta$. So in particular the inconsistent logic is respected by every matrix (for the language)! If there is one and only one matrix **M** respecting L, then we say that **M** is *the matrix implicit in* L.

This new concept motivates the following elaboration of the conventions laid down above. If L is a logic with implicit matrix **M**, then because of its uniqueness, there is no need of explicit reference to **M** in terminology or notation. Thus we shall say that Γ *tautologically implies* Θ ($\Gamma \vDash \Theta$) if $\Gamma \vDash^M \Theta$, and that Γ and Θ are *tautologically equivalent* ($\Gamma \dashv\vDash \Theta$) if $\Gamma \dashv\vDash^M \Theta$. If A is an **M**-tautology or an **M**-contradiction, then we shall say simply that A is a *tautology* or a *contradiction*, respectively.

Yet another convention, which is now open to us and will prove useful, is the following. Let L be a logic with implicit matrix **M**. Let f be any truth-value assignment. Then we will often prefer the new notation

$$\bar{f}A$$

to the more cumbersome $f^M A$. Since **M** is unique, no confusion can result. Restating the definition of implicit matrix, we record the following facts.

COROLLARY 3.2.3. *In a logic with an implicit matrix, if* $\Gamma \vDash \Theta$, *then* $\Gamma \vdash \Theta$.

COROLLARY 3.2.4. *In a logic with an implicit matrix, every tautology is deducible.*

These 'corollaries' look a little like completeness results (the former a strong one, the latter a weak one). However, in the present form they are just terminology and not worth anything. This is because we do not yet know what logics with implicit matrices look like — or even that they exist, for that matter. These are topics that will be discussed in the following sections.

3.3. Types of Boolean operators

We will give necessary and sufficient conditions for a logic to have an implicit matrix in section 3.4. In the present section we introduce an important concept needed for that programme.

Suppose that ☆ is a Boolean operator that directly expresses some truth-value function ϕ under **M**. Suppose that $A_0, \ldots, A_{n-1}$ are any formulas. Let f be any assignment of truth-values to the set of Boolean atoms. Then $f^M(A_0), \ldots, f^M(A_{n-1})$ are uniquely defined quantities. Let

$$I = \{i < n: f^M(A_i) = 1\}, \text{ and } J = \{i < n: f^M(A_i) = 0\}.$$

Then it is easy to see that one of the following two cases holds: either $f^M(\star[A_0, \ldots, A_{n-1}]) = 0$ and

(0) $\qquad \star[A_0, \ldots, A_{n-1}], \{A_i: i \in I\} \vDash^M \{A_i: i \in J\},$

or else $f^M(\star[A_0, \ldots, A_{n-1}]) = 1$ and

(1) $\qquad \{A_i: i \in I\} \vDash^M \{A_i: i \in J\}, \star[A_0, \ldots, A_{n-1}].$

In other words,

$$f^M(\star[A_0, \ldots, A_{n-1}]) = 0 \text{ iff } (0),$$
$$f^M(\star[A_0, \ldots, A_{n-1}]) = 1 \text{ iff } (1).$$

We now introduce some technical definitions. If n is any natural number, then we say that $\langle I, J \rangle$ is a *partitioning of* n if I and J are sets of natural numbers such that

$$I \cup J = \{0, 1, \ldots, n-1\}, \text{ and } I \cap J = \emptyset.$$

Thus, in particular, $\langle I, J \rangle$ is a partitioning of 0 if and only if $I = J = \emptyset$.

Suppose that L is a logic and that $\langle I, J \rangle$ is a partitioning of some n. Then we say that an n-ary operator $\star$ is *of type* 0 *with respect to* $\langle I, J \rangle$ (*in* L) if, for all $A_0, \ldots, A_{n-1}$,

(0') $\qquad \star[A_0, \ldots, A_{n-1}], \{A_i: i \in I\} \vdash \{A_i: i \in J\},$

while we say that $\star$ is *of type* 1 *with respect to* $\langle I, J \rangle$ (*in* L), if for all $A_0, \ldots, A_{n-1}$,

(1') $\qquad \{A_i: i \in I\} \vdash \{A_i: i \in J\}, \star[A_0, \ldots, A_{n-1}].$

It is readily seen that if L is closed under substitution, then $\star$ is of type 0 with respect to $\langle I, J \rangle$ if and only if, for any distinct propositional letters $P_0, \ldots, P_{n-1}$,

(0") $\qquad \star[P_0, \ldots, P_{n-1}], \{P_i: i \in I\} \vdash \{P_i: i \in J\},$

and that $\star$ is of type 1 with respect to $\langle I, J \rangle$ if and only if, for any distinct propositional letters $P_0, \ldots, P_{n-1}$,

(1") $\qquad \{P_i: i \in I\} \vdash \{P_i: i \in J\}, \star[P_0, \ldots, P_{n-1}].$

The similarity between the semantic conditions (0), (1) above and the syntactic conditions (0'), (1') and (0"), (1") is obvious.

If the notion of type determination seems complicated or obscure, then let it be remarked that, essentially, each condition (0") or (1") corresponds to a row in a truth-table. For suppose that ☆ is an n-ary Boolean operator, and that $P_0, \ldots, P_{n-1}$ are distinct propositional letters. Consider the truth-table of ☆ (under some matrix that respects the given logic) with $P_0, \ldots, P_{n-1}$ as the independent variables. Suppose that, in a given row of the truth-table, $P_{i_0}, \ldots, P_{i_{p-1}}$ are the propositional letters that are given the truth-value 1 (true) and that $P_{j_0}, \ldots, P_{j_{q-1}}$ are the ones given the truth-value 0 (false). Let $I = \{i_0, \ldots, i_{p-1}\}$, and $J = \{j_0, \ldots, j_{q-1}\}$. Then obviously $\langle I, J \rangle$ is a partitioning of n, and the truth-value of ☆$[P_0, \ldots, P_{n-1}]$ in this particular row is 1 or 0 depending on whether ☆ is of type 1 or type 0 with respect to $\langle I, J \rangle$. This intuitive remark is made precise in the following lemma:

LEMMA 3.3.1. *Let* L *be a common logic and* **M** *a matrix that respects* L. *Let* ☆ *be an* n-*ary Boolean operator. Suppose that* $x_0, \ldots, x_{n-1} \in \{0, 1\}$, *and let* $I = \{i < n : x_i = 1\}$, *and* $J = \{i < n : x_i = 0\}$. *Then* ☆ *is of type* 0 *(or* 1*) with respect to* $\langle I, J \rangle$ *if and only if* $\mathbf{M}(☆)(x_0, \ldots, x_{n-1}) = 0$ *(respectively*, 1*).*

Proof. Let $P_0, \ldots, P_{n-1}$ be any n distinct propositional letters. (This assumption is acceptable since, by general assumption, there are infinitely many propositional letters.) Let f be any truth-value assignment such that $fP_i = x_i$, for all $i < n$. Note that

$$f^{\mathbf{M}}(☆[P_0, \ldots, P_{n-1}]) = \mathbf{M}(☆)(x_0, \ldots, x_{n-1}).$$

The following chain of reasoning now goes through:

☆ is of type 0 with respect to $\langle I, J \rangle$

iff ☆$[P_0, \ldots, P_{n-1}], \{P_i : i \in I\} \vDash^{\mathbf{M}} \{P_i : i \in J\}$;

iff $f^{\mathbf{M}}(☆[P_0, \ldots, P_{n-1}]) = 0$;

iff $\mathbf{M}(☆)(x_0, \ldots, x_{n-1}) = 0$.

For the type 1 case there is an analogous argument. ∎

Against this intuitive background it is of course important that the following result holds:

THEOREM 3.3.2. *Let* L *be a consistent, assertive, substitutional cut logic that satisfies conditions* (0″) *and* (1″). *Then no Boolean operator* ☆ *is of both type* 0 *and type* 1 *with respect to the same partitioning.*

Proof. Let ☆ be an n-ary Boolean operator. By way of contradiction, suppose that ☆ is of both types with respect to the same $\langle I, J \rangle$. Let $P_0, \ldots, P_{n-1}$ be distinct propositional letters. Then (0″) and (1″) hold. Since L is closed under cut, it follows that

(†) $\qquad \{P_i : i \in I\} \vdash \{P_i : i \in J\}.$

There are three cases.

Case A. $I = \emptyset$. In this case (†) yields $\vdash \{P_i : i \in J\}$. Hence, by substitutivity, $\vdash A$, for every formula A. But Form $\vdash$, by assertivity. Hence, by cut, $\vdash$, which contradicts consistency.

Case B. $J = \emptyset$. This case is similar to Case A.

Case C. $I \cup J \neq \emptyset$. Then, by (†) and substitutivity, $A \vdash B$, for all A, B. L is left-assertive, so by Theorem 2.3.4(i), $A \vdash$, for all A. But L is right-assertive as well, so by Theorem 2.3.4(ii), $\vdash$, which contradicts consistency. ∎

Let us say that an n-ary Boolean operator ☆ is *type determined* (*in* L) if, for every partitioning $\langle I, J \rangle$ of n, ☆ is of either type 0 or type 1 with respect to $\langle I, J \rangle$. The concept of type determination will play an important role in what follows. Actually it would be possible to generalize this notion and speak of type determination in connection not only with Boolean operators but with purely Boolean formulas in general. For example, it is easy to prove the following result.

THEOREM 3.3.3. *Let* L *be a consistent, assertive, substitutional cut logic in which every Boolean operator is type determined. Suppose that* A *is a Boolean compound of some distinct propositional letters* $P_0, \ldots, P_{n-1}$. *Let* $\langle I, J \rangle$ *be any partitioning of* n. *Then the conditions*

(0*) $\qquad A, \{P_i : i \in I\} \vdash \{P_i : i \in J\},$

(1*) $\qquad \{P_i : i \in I\} \vdash \{P_i : i \in J\}, A$

cannot both hold. On the other hand, if L *has an implicit matrix, then at least one of them does.*

However, such a generalized notion is not really needed here, and so we abstain from introducing it formally.

3.4. A completeness theorem in Boolean logic

We shall now show that, for a consistent, assertive, common logic all of whose Boolean operators are type determined, there is one and only one matrix respecting it. This is the main result of this chapter. As the proof is rather long, we separate the existence and uniqueness parts, beginning with the easier one, the latter.

THEOREM 3.4.1. *Let* L *be a consistent common logic. Suppose that* L *is assertive. Then there is at most one matrix that respects* L.

Proof. Let L be as stated. Suppose that M_1 and M_2 are matrices respecting L. Assume that $M_1 \neq M_2$. Then there must be some Boolean operator ☆, n-ary, say, such that, for some $x_0, \ldots, x_{n-1} \in \{0, 1\}$,

$$M_1(☆)(x_0, \ldots, x_{n-1}) \neq M_2(☆)(x_0, \ldots, x_{n-1}).$$

Let $I = \{i : x_i = 1\}$, and $J = \{i : x_i = 0\}$. Then $\langle I, J \rangle$ is a partitioning of n. By Lemma 3.3.1, ☆ is of both type 0 and type 1 with respect to $\langle I, J \rangle$. Since L is consistent, assertive, and common, Theorem 3.3.2 applies, and so we have a contradiction. ∎

THEOREM 3.4.2. *Let L be an assertive common logic. Suppose that every Boolean operator is type determined in L. Then there is a matrix that respects L.*

Proof. If L is inconsistent the situation is trivial, so we may as well assume that L is consistent. On this assumption we shall define a matrix M which will be shown to respect L. For any n-ary Boolean operator ☆ and any $x_0, \ldots, x_{n-1} \in \{0, 1\}$ we define

$$M(☆)(x_0, \ldots, x_{n-1}) = 1, \text{ if ☆ is of type 1 with respect to } \langle I, J \rangle,$$

$$= 0, \text{ if ☆ is of type 0 with respect to } \langle I, J \rangle,$$

where $I = \{i < n : x_i = 1\}$, and $J = \{i < n : x_i = 0\}$. Since L is consistent, assertive and common, Theorem 3.3.2 applies, so the definition of M is correct. Moreover, since every Boolean operator ☆ is type determined, $M(☆)$ is everywhere defined.

Having defined M we must now prove that M respects L. To do this we will use one of the Lindenbaum lemmas above, viz. Theorem 2.6.5. However, that result applies only to finitary logics (closed under monotonicity and cut), and here we have not assumed that L is finitary. We shall circumvent this difficulty by proving a slightly stronger result: There is a finitary, reflexive, monotonic logic S closed under cut such that S is a sublogic of L, and M respects S; hence, *a fortiori*, M respects L.

Since every Boolean operator is type determined in L, there are a number of instances of the conditions (0′) and (1′) in Section 3.3 that hold in L. Let S_0 be the set of pairs $\langle \Gamma, \Theta \rangle$ that are described by all of these instances. Hence — obvious but crucial — every member of S_0 is a pair of finite sets of formulas.

If Form is the set of formulas in the given language, we define

$$S_1 = S_0 \cup \{\langle\{A\}, \{A\}\rangle : A \in \text{Form}\}.$$

Note that every member of S_1, too, is a pair of finite sets of formulas. We will define S_2 as the closure of S_1 under (Cut_G). This may be accomplished by defining an infinite family of sets S_1^n as follows:

$S_1^0 = S_1$,

$S_1^{n+1} = S_1^n \cup \{\langle \Gamma \cup \Gamma', \Theta \cup \Theta' \rangle : \exists A (\Gamma \vdash \Theta, A \text{ and } A, \Gamma' \vdash \Theta' \text{ in } S_1^n)\}$.

$S_2 = \bigcup_{n<\omega} S_1^n$.

Hence, also every member of S_2 is a pair of finite sets of formulas. Finally, let S_3 be the closure of S_2 under monotonicity. Then, obviously, S_3 is a finitary logic. That S_3 is closed under cut requires more of an argument.

By Theorem 2.3.6 it is enough to show that S_3 is closed under (Cut$_G$). Therefore, assume that, for some A and some $\Gamma, \Gamma', \Theta, \Theta', \Gamma \vdash_{S_3} \Theta, A$, and $A, \Gamma' \vdash_{S_3} \Theta'$. By the definition of S_3 and the nature of S_2, there will then be finite sets $\Gamma_0 \subseteq \Gamma, \Gamma_0' \subseteq \Gamma', \Theta_0 \subseteq \Theta, \Theta_0' \subseteq \Theta'$ such that $\Gamma_0 \vdash_{S_2} \Theta_0, (A)$, and $(A), \Gamma_0' \vdash_{S_2} \Theta_0'$, where the parentheses around A indicate that, possibly, the condition holds with A removed from it. If either $\Gamma_0 \vdash_{S_2} \Theta_0$ or $\Gamma_0' \vdash_{S_2} \Theta_0'$, then, since $S_2 \subseteq S_3$ and S_3 is monotonic, $\Gamma, \Gamma' \vdash_{S_3} \Theta, \Theta'$ at once follows. So assume therefore that $\Gamma_0 \vdash_{S_2} \Theta_0, A$, and $A, \Gamma_0' \vdash_{S_2} \Theta_0'$. Since S_2 is closed under (Cut$_G$), $\Gamma_0, \Gamma_0' \vdash_{S_2} \Theta_0, \Theta_0'$, where, again by the fact that $S_2 \subseteq S_3$ and S_3 is monotonic, $\Gamma, \Gamma' \vdash_{S_3} \Theta, \Theta'$. Hence S_3 is indeed closed under cut.

Since $S_0 \subseteq L$ and L is closed under all the closure conditions mentioned, it follows that also $S_3 \subseteq L$. Hence S_3 can be taken as the finitary, reflexive, monotonic sublogic of L closed under cut that we set out to find. Accordingly, for the remainder of the proof, we will write just S for S_3.

Assume now that Γ and Θ are particular sets of formulas, not necessarily finite, such that $\Gamma \not\vdash_L \Theta$. Since $S \subseteq L$, it follows that $\Gamma \not\vdash_S \Theta$. But S is finitary as well as closed under monotonicity and cut; hence, we may apply Theorem 2.6.5: there are $\Gamma^* \supseteq \Gamma$ and $\Theta^* \supseteq \Theta$ such that $\Gamma^* \not\vdash_S \Theta^*$ and $\Gamma^* \cup \Theta^* =$ Form. (Note that here we indirectly invoke the Axiom of Choice.)

Let f be the truth-value assignment for the set of Boolean atoms such that

$fA = 1$, if $A \in \Gamma^*$,

$= 0$, if $A \in \Theta^*$.

Notice that this definition of f is correct, for, as $\Gamma^* \not\vdash_S \Theta^*$, the fact that S is reflexive implies that $\Gamma^* \cap \Theta^* = \emptyset$. (This is precisely why we elected to make S reflexive!)

We contend that f extends to f^M in such a way that

$f^M A = 1$, if $A \in \Gamma^*$,

$= 0$, if $A \in \Theta^*$.

This contention is proved by an inductive argument, the basis of which is given by the definition of f. Consider any formula $\star[A_0, \ldots, A_{n-1}]$ where $\star$ is an n-ary Boolean operator and the contention is known to hold for $A_0, \ldots, A_{n-1}$.

Suppose that $I = \{i < n : A_i \in \Gamma^*\}$, and $J = \{i < n : A_i \in \Theta^*\}$. Note that since $\Gamma^* \cup \Theta^* = $ Form, $\langle I, J \rangle$ is a partitioning of n. By the induction hypothesis, $I = \{i < n : f^M A_i = 1\}$, and $J = \{i < n : f^M A_i = 0\}$. Hence, by the definition of **M** and the fact that $f^M(\star[A_0, \ldots, A_{n-1}]) = M(\star)(f^M A_0, \ldots, f^M A_{n-1})$,

$$f^M(\star[A_0, \ldots, A_{n-1}]) = 1, \text{ if } \star \text{ is of type 1 with respect to } \langle I, J \rangle,$$
$$= 0, \text{ if } \star \text{ is of type 0 with respect to } \langle I, J \rangle.$$

Suppose first that $f^M(\star[A_0, \ldots, A_{n-1}]) = 0$. Then $\star$ is of type 0 with respect to $\langle I, J \rangle$, and so

$$\star[A_0, \ldots, A_{n-1}], \{A_i : i \in I\} \vdash_S \{A_i : i \in J\}.$$

Now $\{A_i : i \in I\} \subseteq \Gamma^*$ and $\{A_i : i \in J\} \subseteq \Theta^*$. If it were the case that $\star[A_0, \ldots, A_{n-1}] \in \Gamma^*$, then monotonicity would yield $\Gamma^* \vdash_S \Theta^*$, which is impossible. Therefore instead $\star[A_0, \ldots, A_{n-1}] \in \Theta^*$. By an analogous argument it can be shown that $f^M(\star[A_0, \ldots, A_{n-1}]) = 1$ only if $\star[A_0, \ldots, A_{n-1}] \in \Gamma^*$. This completes the induction.

Now we have arrived at the goal. The existence of f^M shows that $\Gamma^* \nvdash \Theta^*$. *A fortiori*, then, $\Gamma \nvdash \Theta$. So **M** respects L. ∎

The last two theorems yield what we have announced as the main result of this chapter.

THEOREM 3.4.3. *A necessary and sufficient condition for a consistent common logic to possess an implicit matrix is that it is assertive and every Boolean operator is type determined in it.*

The logics singled out in the preceding theorem are sufficiently important to deserve a name of their own. Let us call a common logic *Boolean* if it is assertive and every Boolean operator is type determined in it. In this terminology Theorem 3.4.3 amounts to this:

COROLLARY 3.4.4. *A consistent common logic has an implicit matrix if and only if it is Boolean.*

The reader will recall that we remarked of Corollaries 3.2.3 and 3.2.4 that they looked a little like completeness results. This was because of their verbal similarity to standard completeness theorems; it was emphasized that by themselves they were quite empty. But supplemented with the result just proved, they become (at least partial) completeness results also in substance:

COROLLARY 3.4.5. *In a Boolean logic, if $\Gamma \vDash \Theta$, then $\Gamma \vdash \Theta$.*

COROLLARY 3.4.6. *In a Boolean logic, every tautology is deducible.*

The last few remarks may deserve a comment. The point of completeness theorems is to relate syntactic and semantic concepts. The concepts 'syntactic' and 'semantic' have not been given precise definitions here. But generally speaking, semantics has to do with meaning, while syntax concerns the structure of formulas or sets of formulas. Matrices are called up in order to assign truth-values to Boolean compounds and to that extent may be said to bring out the meaning of the Boolean operators – the meaning of a Boolean operator being taken to be the corresponding truth-value function. Thus 'matrix' is a semantic concept. On the other hand, conditions such as these:

(α) $\quad\quad\quad\quad\quad\quad\quad\quad \Gamma \vdash \Theta$

(β) $\quad\quad\quad\quad\quad\quad\quad\quad \Gamma \vdash \Theta \Rightarrow \Gamma' \vdash \Theta'$

(γ) $\quad\quad\quad\quad\quad\quad\quad\quad \Gamma \vdash \Theta \,\&\, \Gamma' \vdash \Theta' \Rightarrow \Gamma'' \vdash \Theta''$

are syntactic if the descriptions of the sets involved are in terms of formula structure and set structure (that is, in terms of what the formulas and sets 'look like'). One may think of more complicated kinds of conditions that would also be called syntactic. But already on the present vague understanding, the conditions going into the definition of being Boolean are immediately recognized as syntactic. That is to say, Booleanhood belongs to syntax. In this sense there is a clear connection between traditional completeness results and the last two corollaries. At the same time it is important not to claim too much. The 'completeness results' were qualified above as 'partial'. This was in order to stress that the converses of Corollaries 3.4.5 and 3.4.6 are not asserted. Indeed, whenever the language contains non-Boolean operators of an interesting kind, the converses cannot be asserted.

Some other comments should be made before closing this section. The proof of Theorem 3.4.2 belongs to a rich tradition going back to Leon Henkin (see Henkin 1949). His ideas have become standard in the classical analysis of (not necessarily classical) propositional logic (see, for example, Aczel 1967; Cresswell 1967; Fine 1974; Goldblatt 1974; Kaplan 1966; Lemmon 1977; Makinson 1966; Routley and Meyer 1972 a, b, c; Segerberg 1968; R. H. Thomason 1968). Of course, even within this tradition our proof is unusual in that the object language is not exhibited. The concept of type determination – especially in the light of the intuitive remarks preceding Lemma 3.3.1 – may lead the idea to László Kalmár; see the well-known account of his completeness proof for classical propositional logic in Kleene (1952). We shall return briefly to this topic in the following section.

3.5. Congruence in Boolean logic

Are Boolean operators congruential (in Boolean logics)? So one would expect. It is somewhat reassuring that this result is indeed forthcoming. To derive it we need a lemma.

LEMMA 3.5.1. *Let L be a common logic. Suppose that Γ and Θ are sets of formulas and that there are some n and some formulas $C_0, \ldots, C_{n-1}$ such that, for every partitioning $\langle I, J \rangle$ of n,*

$$\Gamma, \{C_i : i \in I\} \vdash \{C_i : i \in J\}, \Theta.$$

Then $\Gamma \vdash \Theta.$

Proof. We proceed by induction on n. If n = 0, the result holds vacuously. Next suppose, as our induction hypothesis, that the lemma holds for n. Let $C_0, \ldots, C_{n-1}, C_n$ be formulas such that, for every partitioning $\langle I, J \rangle$ of n + 1,

(1) $\qquad \Gamma, \{C_i : i \in I\} \vdash \{C_i : i \in J\}, \Theta.$

Let $\langle I^\circ, J^\circ \rangle$ be any partitioning of n. Then both $\langle I^\circ \cup \{n\}, J^\circ \rangle$ and $\langle I^\circ, J^\circ \cup \{n\} \rangle$ are partitionings of n + 1. Hence, by (1),

(2) $\qquad \Gamma, C_n, \{C_i : i \in I^\circ\} \vdash \{C_i : i \in J^\circ\}, \Theta,$

(3) $\qquad \Gamma, \{C_i : i \in I^\circ\} \vdash \{C_i : i \in J^\circ\}, C_n, \Theta.$

By (2), (3), and cut, then,

$$\Gamma, \{C_i : i \in I^\circ\} \vdash \{C_i : i \in J^\circ\}, \Theta.$$

From the induction hypothesis it now follows that $\Gamma \vdash \Theta.$ ∎

THEOREM 3.5.2. *Let L be a Boolean logic. Then every Boolean operator is congruential in L.*

Proof. Let ☆ be an n-ary Boolean operator, and suppose that $A_0, \ldots, A_{n-1}$, $B_0, \ldots, B_{n-1}$ are formulas such that

(1) $\qquad A_0 \dashv\vdash B_0, \ldots, A_{n-1} \dashv\vdash B_{n-1}.$

Let U_0 and U_1 be the sets of partitionings of n with respect to which ☆ is of type 0 respectively type 1. Then, for all $\langle I, J \rangle \in U_0$,

(2) $\qquad \text{☆}[A_0, \ldots, A_{n-1}], \{A_i : i \in I\} \vdash \{A_i : i \in J\},$

while, for all $\langle I, J \rangle \in U_1$,

(3) $\qquad \{B_i : i \in I\} \vdash \{B_i : i \in J\}, \text{☆}[B_0, \ldots, B_{n-1}].$

We claim that, for all partitionings $\langle I, J \rangle$ of n,

(†) $\qquad \text{☆}[A_0, \ldots, A_{n-1}], \{A_i : i \in I\} \vdash \{A_i : i \in J\}, \text{☆}[B_0, \ldots, B_{n-1}].$

By (2) and monotonicity, it is clear that (†) holds for all $\langle I, J \rangle \in U_0$. Assume therefore that $\langle I, J \rangle$ is a particular partitioning of n such that $\langle I, J \rangle \notin U_0$. Since L is Boolean, ☆ is type determined. Therefore $\langle I, J \rangle \in U_1$. By (1) and monotonicity,

(4) $\qquad \{A_i : i \in I\} \vdash B_k$, for every $k \in I$;

(5) $\qquad B_k \vdash \{A_i : i \in J\}$, for every $k \in J$.

Hence, by (3), (4), and cut,

(6) $\qquad \{A_i : i \in I\} \vdash \{B_i : i \in J\}, ☆[B_0,\ldots, B_{n-1}]$,

and by (5), (6), and cut

(7) $\qquad \{A_i : i \in I\} \vdash \{A_i : i \in J\}, ☆[B_0,\ldots, B_{n-1}]$,

from which (†) follows by monotonicity. Hence (†) holds for all partitionings $\langle I, J \rangle$ of n. Consequently, by Lemma 3.5.1,

$$☆[A_0,\ldots, A_{n-1}] \vdash ☆[B_0,\ldots, B_{n-1}].$$

A completely symmetric argument yields the converse:

$$☆[B_0,\ldots, B_{n-1}] \vdash ☆[A_0,\ldots, A_{n-1}]. \blacksquare$$

Now we are in a position to return to Kalmár's completness proof, which was mentioned in the preceding section, and give a generalized account of it. The result to be proved, Theorem 3.5.3, is weaker than the completeness result proved by Henkin's method, Theorem 3.4.2. But while the Henkin-type proof relied on the Axiom of Choice, the proof that follows is constructive and so in a sense more informative. The situation well illustrates the trade-off between generality and constructivity.

To avoid every chance of misunderstanding, let us repeat that what now follows is a new proof of something we already know to be true. Of course, the new proof does not rely on any non-constructive part of the old proof — if it did, it would not be worth giving here.

THEOREM 3.5.3. *Let L be a Boolean logic. Then there is a matrix* **M** *such that* $A \vDash^{\mathbf{M}}$ *implies* $A \vdash$, *and* $\vDash^{\mathbf{M}} B$ *implies* $\vdash B$, *for all formulas* A, B.

Proof. Assume that L is consistent (otherwise the proof is trivial). Let **M** be the matrix defined in the proof of Theorem 3.4.2; again, since L is consistent and Boolean, **M** exists, and every **M**(☆) is everywhere defined. Now, with the Henkin method we showed that **M** respects L. Here we will substitute what might be called a Kalmár method and prove the weaker conclusion of the present theorem (that **M** *respects* L *formulawise*, say).

Suppose that $A \vDash^{\mathbf{M}}$. If A were a Boolean atom, then $A \nvDash^{\mathbf{M}}$, contrary to our assumption. Hence $A = ☆[C_0,\ldots, C_{n-1}]$, for some n-ary Boolean operator ☆ and some formulas $C_0, \ldots, C_{n-1}$. Let $\langle I, J \rangle$ be an arbitrary partitioning of n. Then our assumption implies — $\vDash^M$ is monotonic (Theorem 3.2.1) — that

$$\star[C_0, \ldots, C_{n-1}], \{C_i : i \in I\} \vDash^M \{C_i : i \in J\}.$$

So if $x_0, \ldots, x_{n-1} \in \{0, 1\}$ are such that $I = \{i : x_i = 1\}$ and $J = \{i : x_i = 0\}$, then $\mathbf{M}(\star)(x_0, \ldots, x_{n-1}) = 0$. By the definition of $\mathbf{M}$, therefore, $\star$ is of type 0 with respect to $\langle I, J \rangle$. Consequently, since by hypothesis $\star$ is type determined in L,

$$\star[C_0, \ldots, C_{n-1}], \{C_i : i \in I\} \vdash \{C_i : i \in J\}.$$

But this reasoning holds for every partitioning $\langle I, J \rangle$ of n! By Lemma 3.5.1, then, $A \vdash$. The other half of the theorem is proved similarly. ∎

Theorem 3.5.3 may not seem like much compared to Theorem 3.4.2. Our very weak assumptions on the language are partly to blame for this; one section hence it would be possible to do a little better. There it would be possible to prove, by the Kalmár method, that if a Boolean logic satisfies some very mild conditions, then there is a matrix *finitely respecting* the logic, in the sense that whenever Γ and Θ are finite sets of formulas, then $\Gamma \vDash \Theta$ implies $\Gamma \vdash \Theta$. However, we shall be content with merely mentioning this possibility.

3.6. Particular truth-value functions: The Big Seven

Up till now we have not named any particular truth-value functions. Now we will. There are exactly two zeroary truth-value functions, *constant truth*, which we denote by T, and *constant falsity*, which we denote by F. Among the four unary truth-value functions *negation* stands out, viz., the function N such that

$$N(x) = 1 \quad \text{iff} \quad x = 0.$$

Of the eight binary truth-value functions we mention *conjunction*, K; *disjunction*, A; *(material) implication*, C; and *(material) equivalence*, E. We may cast the traditional truth-tables in the following form:

$$K(x, y) = 1 \quad \text{iff} \quad x = y = 1;$$

$$A(x, y) = 1 \quad \text{iff} \quad x = 1 \text{ or } y = 1;$$

$$C(x, y) = 1 \quad \text{iff} \quad x \leq y;$$

$$E(x, y) = 1 \quad \text{iff} \quad x = y;$$

for all truth values x, y. The reader may easily give the conditions for equality to 0 by negating the right-hand sides of these biconditionals. These — T, F, N, K, A, C, E — are the best known truth-value functions in the literature. It is to them we refer as *The Big Seven*.

It is a well-known fact that in classical propositional logic there exist finite — in fact, very small — sets of truth-value functions such that every truth-value function can be generated from the members of such a set by means of functional composition. Examples of such sets are {*F, C*}, {*N, C*}, {*N, K*}, and {*N, A*}. Even though we mention it without proof, this is an important fact, and we will refer to it later.

We shall now consider a number of conditions that can be meaningfully imposed on an operator ⊕ (zeroary, unary, or binary, as the context makes clear). They have been collected in two tables, for obvious reasons called *elimination* conditions (Table 1) and *introduction* conditions (Table 2), respectively. All except the zeroary ones come in pairs with one *antecedent* condition corresponding to one *consequent* condition. (We have defined no concepts of 'antecedent' and 'consequent', so the terminology is to be understood heuristically.) Finally, each condition is something to do with one of The Big Seven; exactly how is made clear below. To refer to these conditions we use a code consisting of three letters. The encoding system should be self-explanatory. Thus **(EAN)** concerns *E*limination in the *A*ntecedent and has to do with negation, *N*, while **(ICC)** concerns *I*ntroduction in the *C*onsequent and has to do with material implication, *C*.

Some simple relations between conditions are worth noting. We have already remarked that most come in pairs. Furthermore, except for the conditions concerning zeroary operators, corresponding EA- and IA-conditions are converses of one another, as are corresponding EC- and IC- conditions. Finally, let us say that EA- and IC-conditions, as well as corresponding EC- and IA-conditions, are *duals* of one another. At first blush, it might seem impossible that an elimination condition could be equivalent to an introduction condition. But in common logics it happens all the time:

THEOREM 3.6.1. *In a common logic, dual conditions are equivalent.*

Proof. What is claimed is that the following equivalences hold:

(i) **(EAT)** iff **(ICT)**;
(ii) **(IAF)** iff **(ECF)**;
(iii) **(EAN)** iff **(ICN)**;
(iv) **(IAN)** iff **(ECN)**;
(v) **(EAK)** iff **(ICK)**;
(vi) **(IAK)** iff **(ECK)**;
(vii) **(EAA)** iff **(ICA)**;
(viii) **(IAA)** iff **(ECA)**;
(ix) **(EAC)** iff **(ICC)**;
(x) **(IAC)** iff **(ECC)**;
(xi) **(EAE)** iff **(ICE)**;
(xii) **(IAE)** iff **(ECE)**.

In all, there are 24 implications to establish, none difficult. Actually they fall into three groups. Several implications presuppose reflexivity, monotonicity, and closure under cut. But for the majority, reflexivity and closure under cut suffice. And for two of them (**(ICT)** implying **(EAT)**, and **(IAF)** implying **(ECF)**) closure under cut is enough by itself.

Table 1
Elimination conditions

EAT	If then	$\oplus, \Gamma \vdash \Theta,$ $\Gamma \vdash \Theta.$	**ECF**	If then	$\Gamma \vdash \Theta, \oplus$ $\Gamma \vdash \Theta.$
EAN	If then	$\oplus A, \Gamma \vdash \Theta,$ $\Gamma \vdash \Theta, A.$	**ECN**	If then	$\Gamma \vdash \Theta, \oplus A,$ $A, \Gamma \vdash \Theta.$
EAK	If then	$A \oplus B, \Gamma \vdash \Theta,$ $A, B, \Gamma \vdash \Theta.$	**ECK**	If then	$\Gamma \vdash \Theta, A \oplus B,$ $\Gamma \vdash \Theta, A$ *and* $\Gamma \vdash \Theta, B.$
EAA	If then	$A \oplus B, \Gamma \vdash \Theta,$ $A, \Gamma \vdash \Theta$ *and* $B, \Gamma \vdash \Theta.$	**ECA**	If then	$\Gamma \vdash \Theta, A \oplus B,$ $\Gamma \vdash \Theta, A, B.$
EAC	If then	$A \oplus B, \Gamma \vdash \Theta,$ $\Gamma \vdash \Theta, A$ *and* $B, \Gamma \vdash \Theta.$	**ECC**	If then	$\Gamma \vdash \Theta, A \oplus B,$ $A, \Gamma \vdash \Theta, B.$
EAE	If then	$A \oplus B, \Gamma \vdash \Theta,$ $A, B, \Gamma \vdash \Theta$ *and* $\Gamma \vdash \Theta, A, B.$	**ECE**	If then	$\Gamma \vdash \Theta, A \oplus B,$ $A, \Gamma \vdash \Theta, B$ *and* $B, \Gamma \vdash \Theta, A.$

Here we will prove only one of the implications, viz., that **(ICK)** implies **(EAK)**. Thus let L be a common logic and $\oplus$ a binary operator satisfying condition **(ICK)**. Suppose that

(1) $\qquad\qquad\qquad A \oplus B, \Gamma \vdash \Theta.$

By reflexivity of L, $A \vdash A$. Hence, by monotonicity of L, $A, B \vdash A$. Similarly, $A, B \vdash B$. Consequently, by **(ICK)**,

(2) $\qquad\qquad\qquad A, B \vdash A \oplus B.$

L is closed under cut, so from (1) and (2) we conclude that

$$A, B, \Gamma \vdash \Theta.$$

Hence $\oplus$ satisfies **(EAK)**. ∎

We have already spoken, a bit loosely, of the conditions of Tables 1 and 2 as 'having something to do' with The Big Seven. Let us now adopt a more precise terminology and say that **(EAT)** and **(ICT)** are the conditions *associated with T*; **(ECF)** and **(IAF)** those *associated with F*; **(EAN)**, **(ECN)**, **(IAN)**, and **(ICN)** those *associated with N*; and so on. Here and below we tacitly assume that when it is asked whether a certain operator satisfies one of these conditions, then it has the appropriate arity (zeroary in the case of the conditions associated with *T* and *F*, unary in the case of *N*, and binary in the case of *K, A, C,* and *E*). The following two theorems explain our choice of terminology:

Table 2
Introduction conditions

	IAF	$\oplus \vdash.$		ICT	$\vdash \oplus.$
IAN	*If then*	$\Gamma \vdash \Theta, A,$ $\oplus A, \Gamma \vdash \Theta.$	ICN	*If then*	$A, \Gamma \vdash \Theta,$ $\Gamma \vdash \Theta, \oplus A.$
IAK	*If then*	$A, B, \Gamma \vdash \Theta,$ $A \oplus B, \Gamma \vdash \Theta.$	ICK	*If then*	$\Gamma \vdash \Theta, A$ *and* $\Gamma \vdash \Theta, B,$ $\Gamma \vdash \Theta, A \oplus B.$
IAA	*If then*	$A, \Gamma \vdash \Theta$ *and* $B, \Gamma \vdash \Theta,$ $A \oplus B, \Gamma \vdash \Theta.$	ICA	*If then*	$\Gamma \vdash \Theta, A, B,$ $\Gamma \vdash \Theta, A \oplus B.$
IAC	*If then*	$\Gamma \vdash \Theta, A$ *and* $B, \Gamma \vdash \Theta,$ $A \oplus B, \Gamma \vdash \Theta.$	ICC	*If then*	$A, \Gamma \vdash \Theta, B,$ $\Gamma \vdash \Theta, A \oplus B.$
IAE	*If then*	$A, B, \Gamma \vdash \Theta$ *and* $\Gamma \vdash \Theta, A, B,$ $A \oplus B, \Gamma \vdash \Theta.$	ICE	*If then*	$A, \Gamma \vdash \Theta, B$ *and* $B, \Gamma \vdash \Theta, A,$ $\Gamma \vdash \Theta, A \oplus B.$

THEOREM 3.6.2. *Let* L *be a consistent Boolean logic. Suppose that* ☆ *is a Boolean operator that directly expresses one of the truth-value functions T, F, N, K, A, C, and E. Then* ☆ *satisfies the associated conditions.*

Proof. There are many cases to check, but the procedure is the same in all of them. We shall give one example. Suppose that ☆ is a binary operator directly expressing E in L. We must check that **(EAE)**, **(ECE)**, **(IAE)**, and **(ICE)** hold.

Since ☆ directly expresses E, we have

$$A, B \vDash A \,☆\, B;$$
$$A, A \,☆\, B \vDash B;$$
$$B, A \,☆\, B \vDash A;$$
$$\vDash A \,☆\, B, A, B.$$

Hence, since the implicit matrix respects L,

(1) $\qquad A, B \vdash A \,☆\, B;$
(2) $\qquad A, A \,☆\, B \vdash B;$
(3) $\qquad B, A \,☆\, B \vdash A;$
(4) $\qquad \vdash A \,☆\, B, A, B.$

Assume now that $A \star B, \Gamma \vdash \Theta$. By (1) and cut, $A, B, \Gamma \vdash \Theta$. By (4) and cut, $\Gamma \vdash \Theta, A, B$. Hence (**EAE**) holds.

Assume instead that $\Gamma \vdash \Theta, A \star B$. By (2) and cut, $A, \Gamma \vdash \Theta, B$. By (3) and cut, $B, \Gamma \vdash \Theta, A$. Hence (**ECE**) holds.

Next assume that

(5) $\qquad\qquad A, B, \Gamma \vdash \Theta;$

(6) $\qquad\qquad \Gamma \vdash \Theta, A, B.$

By (2), (5), and cut, $A, A \star B, \Gamma \vdash \Theta$. By (3), (6), and cut, $A \star B, \Gamma \vdash \Theta, A$. Therefore, by cut, $A \star B, \Gamma \vdash \Theta$. Hence (**IAE**) holds.

Finally assume that

(7) $\qquad\qquad A, \Gamma \vdash \Theta, B;$

(8) $\qquad\qquad B, \Gamma \vdash \Theta, A.$

By (1), (7), and cut, $A, \Gamma \vdash \Theta, A \star B$. By (4), (8), and cut, $\Gamma \vdash \Theta, A \star B, A$. Therefore, by cut, $\Gamma \vdash \Theta, A \star B$. Hence (**ICE**) holds. ∎

THEOREM 3.6.3. *Let L be a consistent Boolean logic. Suppose that $\star$ is a Boolean operator that satisfies the conditions associated with one of the truth-value functions T, F, N, K, A, C, and E. Then $\star$ directly expresses that truth-value function.*

Proof. As an example, suppose that $\star$ is a binary Boolean operator that satisfies the conditions associated with K. Let **M** be the matrix implicit in L. We must show that $\mathbf{M}(\star) = K$.

First note that if P and Q are any distinct fixed propositional letters, then

(1) $\qquad\qquad P, Q \vdash P \star Q.$

To see this, note that $P \vdash P$ and $Q \vdash Q$, by reflexivity, and that hence $P, Q \vdash P$ and $P, Q \vdash Q$, by monotonicity. But then (1) follows by (**ICK**), which $\star$ is supposed to satisfy.

Since $\star$ is a Boolean operator and L is a Boolean logic, $\star$ is type determined. Furthermore, $\langle \{0, 1\}, \emptyset \rangle$ is a partitioning of 2. Consequently,

(2) $\qquad$ either $P, Q \vDash P \star Q$ or $P \star Q, P, Q \vDash.$

Assume, by way of contradiction, that

(3) $\qquad\qquad P \star Q, P, Q \vDash.$

By respect, then,

$$P \star Q, P, Q \vdash.$$

Hence, by (1), (3), and cut, $P, Q \vdash$. Or, for simplicity,

(4) $\quad\quad\quad\quad\quad\quad\quad\quad P \vdash,$

since L is closed under substitution. By assertivity there is some set Δ such that

(5) $\quad\quad\quad\quad\quad\quad\quad\quad \vdash \Delta.$

But by (4) and substitutivity,

(6) $\quad\quad\quad\quad\quad\quad\quad\quad A \vdash,$ for every $A \in \Delta.$

Applying cut to (5) and (6) we deduce $\vdash$, contradicting the consistency of L. This shows that the assumption (3) is absurd. Hence, from (2),

(7) $\quad\quad\quad\quad\quad\quad\quad\quad P, Q \vDash P \star Q.$

In a similar way the fact that $\star$ satisfies (**IAK**) can be used to show that

(8) $\quad\quad\quad\quad\quad\quad\quad\quad P \star Q \vDash P;$

(9) $\quad\quad\quad\quad\quad\quad\quad\quad P \star Q \vDash Q.$

Now we have everything we need. From (7) it follows that $\mathbf{M}(\star)(1, 1) = 1$. From (8) and (9) it follows that $\mathbf{M}(\star)(x, y) = 1$ only if $x = y = 1$. Consequently, $\mathbf{M}(\star) = K$. ∎

The following remarks will put the preceding few theorems into perspective. Any truth-value function is amenable to a treatment similar to that which we have accorded to the Big Seven. To see this, suppose that ϕ is any n-ary truth-value function, and let $\oplus$ be any n-ary Boolean operator.

There will be one elimination condition associated with ϕ for each partitioning $\langle I, J \rangle$ of n. Suppose that $x_0, \ldots, x_{n-1} \in \{0, 1\}$ and that $I = \{i : x_i = 1\}$ and $J = \{i : x_i = 0\}$. Then, if $\phi(x_0, \ldots, x_{n-1}) = 1$, we would define:

(**EA**$\phi_{\langle I, J \rangle}$) $\quad$ If $\quad\quad\quad \oplus[A_0, \ldots, A_{n-1}], \Gamma \vdash \Theta,$
$\quad\quad\quad\quad\quad\quad$ then $\quad \{A_i : i \in I\}, \Gamma \vdash \Theta, \{A_i : i \in J\}.$

On the other hand, if $\phi(x_0, \ldots, x_{n-1}) = 0$, we would define:

(**EC**$\phi_{\langle I, J \rangle}$) $\quad$ If $\quad\quad\quad \Gamma \vdash \Theta, \oplus[A_0, \ldots, A_{n-1}],$
$\quad\quad\quad\quad\quad\quad$ then $\quad \{A_i : i \in I\}, \Gamma \vdash \Theta, \{A_i : i \in J\}.$

There will always be two general introduction conditions associated with ϕ:

(**IA**ϕ)
$\quad\quad\quad$ If, for all partitionings $\langle I, J \rangle$ of n such that,
$\quad\quad\quad$ for all $x_0, \ldots, x_{n-1} \in \{0, 1\}, I = \{i : x_i = 1\}$ and
$\quad\quad\quad J = \{i : x_i = 0\}, \phi(x_0, \ldots, x_{n-1}) = 1$ and
$\quad\quad\quad\quad\quad \{A_i : i \in I\}, \Gamma \vdash \Theta, \{A_i : i \in J\},$
$\quad\quad\quad$ then $\quad\quad \oplus[A_0, \ldots, A_{n-1}], \Gamma \vdash \Theta.$

(ICϕ) If, for all partitionings $\langle I, J \rangle$ of n such that, for all $x_0,\ldots, x_{n-1} \in \{0, 1\}$, $I = \{i : x_i = 1\}$ and $J = \{i : x_i = 0\}$, $\phi(x_0,\ldots, x_{n-1}) = 0$ and
$$\{A_i : i \in I\}, \Gamma \vdash \Theta, \{A_i : i \in J\},$$
then $\Gamma \vdash \Theta, \oplus[A_0,\ldots, A_{n-1}]$.

Constant truth-value functions command special interest. For $t \in \{0, 1\}$, let us say that ϕ *is identically equal to* t, in symbols $\phi \equiv t$, if, for all $x_0,\ldots, x_{n-1} \in \{0, 1\}$, $\phi(x_0,\ldots, x_{n-1}) = t$. Then if $\phi \equiv 0$, (IAϕ) reduces to

(IAF$_n$) $\oplus[A_0,\ldots, A_{n-1}], \Gamma \vdash \Theta$;

while if $\phi \equiv 1$, (ICϕ) reduces to

(ICT$_n$) $\Gamma \vdash \Theta, \oplus[A_0,\ldots, A_{n-1}]$.

In a context where monotonicity is assumed, these conditions reduce even further to

(IAF$'_n$) $\oplus[A_0,\ldots, A_{n-1}] \vdash$;

(ICT$'_n$) $\vdash \oplus[A_0,\ldots, A_{n-1}]$.

There is still a certain symmetry in the picture; for example, (IAϕ) and the conjunction of all (EA$\phi_{\langle I,J \rangle}$) are converses, as are (ICϕ) and the conjunction of all (EC$\phi_{\langle I,J \rangle}$). But in general the situation is more complicated than Tables 1 and 2 may suggest. There are 2^n elimination conditions and only 2 general introduction conditions, and there seems to be no routine for aggregating them into a small number of 'nice' conditions. In particular, there would ordinarily seem to be no hope of finding antecedent and consequent conditions as nicely symmetric as those associated with N, K, A, C, and E.

These remarks also explain the lack of symmetry in the case of T and F. First, these truth-value functions are zeroary, so associated with each is just $2^0 = 1$ elimination condition, viz., (EAT$_{\langle \emptyset,\emptyset \rangle}$) and (ECF$_{\langle \emptyset,\emptyset \rangle}$), which are identical with the familiar conditions (EAT) and (ECF) of Table 1. Second, even though there are two introduction conditions associated with each of T and F, they do not come to much. As was already pointed out, in monotonic logics (IAϕ) reduces to (IAF$'_0$) if $\phi = F$, and similarly (ICϕ) to (ICT$'_0$) if $\phi = T$; and those conditions are identical with the familiar (IAF) and (ICT) of Table 2. On the other hand, by Lemma 3.5.1, (IAϕ) becomes trivial in common logics if $\phi \equiv 1$, and similarly (ICϕ) if $\phi \equiv 0$. Thus in a sense there just are not any conditions that we could label (ECT), (IAT), (EAF), or (ICF).

3.7. Arbitrary truth-value functions

Is it possible to do proof theory in a setting as general as ours? In this section — which can be omitted without loss of continuity — we shall briefly explore this question and show what a Gentzen-type analysis of truth-functional logic in our style might look like.

Fix any language $\mathcal{L} = \langle \text{Lett, Bop, Iop, r}\rangle$, and let **M** be any matrix for it. By Theorem 3.2.1, **M** determines a consistent common logic; call it **L(M)**. Our goal in this section will be to give a (partial) characterization of **L(M)**.

We shall now give a syntactic definition of an ancillary logic to be called $L_{\ddagger}$. Like so much in this book, this definition is technically complicated even though the idea is simple: $L_{\ddagger}$ will be the smallest logic that satisfies the diagonality condition of Section 2.2 and is closed under the introduction rules mentioned at the end of Section 3.6.

For the formal definition it is expedient to fall back on the notion of an inductive system (see Section 0.2). That is to say, we define $L_{\ddagger}$ as the set of objects induced by the inductive system $\langle B_{\ddagger}, R_{\ddagger}\rangle$, where $B_{\ddagger}$ and $R_{\ddagger}$ are defined as follows. The set $B_{\ddagger}$ of *basic objects* is the union $B_{\text{diag}} \cup B_0 \cup B_1$, where

$$B_{\text{diag}} = \{\langle \Gamma, \Theta\rangle : \Gamma \cap \Theta \neq \emptyset\},$$

$$B_0 = \{\langle \Gamma, \Theta\rangle : \exists \oplus, n, A_0, \ldots, A_{n-1}(r(\oplus) = n \ \&$$
$$\mathbf{M}(\oplus) \equiv 0 \ \& \ \oplus[A_0, \ldots, A_{n-1}] \in \Gamma)\},$$

$$B_1 = \{\langle \Gamma, \Theta\rangle : \exists \oplus, n, A_0, \ldots, A_{n-1}(r(\oplus) = n \ \&$$
$$\mathbf{M}(\oplus) \equiv 1 \ \& \ \oplus[A_0, \ldots, A_{n-1}] \in \Theta)\}.$$

(For the definition of $\equiv$, see the end of the preceding section.) The *generating relation* $R_{\ddagger}$ consists of all (q + 1)-tuples $\langle\langle \Gamma_0, \Theta_0\rangle, \ldots, \langle \Gamma_{q-1}, \Theta_{q-1}\rangle, \langle \Gamma^*, \Theta^*\rangle\rangle$ that satisfy the following condition: there is a Boolean operator $\oplus$, n-ary say, some formulas $A_0, \ldots, A_{n-1}$, some formula sets Γ, Θ, and a type $t \in \{0, 1\}$ such that, if $\langle I_0, J_0\rangle, \ldots, \langle I_{q-1}, J_{q-1}\rangle$ are all the partitionings of n with respect to which $\oplus$ is of type t (in **L(M)**), then, for all $k < q$,

(†)
$$\Gamma_k = \Gamma \cup \{A_i : i \in I_k\},$$
$$\Theta_k = \Theta \cup \{A_i : i \in J_k\},$$

and either

($\P_1$)
$$t = 1,$$
$$\Gamma^* = \Gamma \cup \{\oplus[A_0, \ldots, A_{n-1}]\},$$
$$\Theta^* = \Theta,$$
$$\oplus[A_0, \ldots, A_{n-1}] \notin \Gamma,$$

or else
$$t = 0,$$
$$\Gamma^* = \Gamma,$$
($\P_0$)
$$\Theta^* = \Theta \cup \{\oplus[A_0,\ldots, A_{n-1}]\},$$
$$\oplus[A_0,\ldots, A_{n-1}] \notin \Theta.$$

The definition of $\langle B_\ddagger, R_\ddagger \rangle$ does look formidable. Notice, however, that B_{diag} corresponds to the condition (Diag) of Section 2.2; that B_0 and B_1 correspond to conditions $(\mathbf{IAF_n})$ and $(\mathbf{ICT_n})$ of Section 3.6; and that, in a similar fashion, the definition of $R_\ddagger$ corresponds to conditions $(\mathbf{IAM}(\oplus))$ and $(\mathbf{ICM}(\oplus))$ of the latter section. The next observation is important (and helps explain our interest in $L_\ddagger$).

LEMMA 3.7.1. *If* $\langle\langle \Gamma_0, \Theta_0 \rangle, \ldots, \langle \Gamma_{q-1}, \Theta_{q-1} \rangle, \langle \Gamma^*, \Theta^* \rangle\rangle \in R_\ddagger$, *then* $\Gamma^* \vDash^M \Theta^*$ *iff* $\Gamma_k \vDash^M \Theta_k$, *for all* $k < q$.

Proof. Assume the hypothesis of the lemma. Without spelling out every detail, let us remind ourselves of what this means: that there is an n-ary Boolean operator $\oplus$, formulas $A_0, \ldots, A_{n-1}$, sets Γ, Θ of formulas, a type $t \in \{0, 1\}$, and partitionings $\langle I_0, J_0 \rangle, \ldots, \langle I_{q-1}, J_{q-1} \rangle$ of n, all satisfying the conditions stipulated in the definition of $R_\ddagger$ above; that is (†) and ($\P_t$).

First suppose that

($\$$) $$\Gamma^* \vDash^M \Theta^*.$$

Take any $k < q$. It follows from the hypothesis that $\oplus$ is of type t with respect to $\langle I_k, J_k \rangle$. Suppose now, for a *reductio ad absurdum*, that $\Gamma_k \nvDash^M \Theta_k$. Then there must be a truth-value assignment f for the set of Boolean atoms such that, for all $A \in \Gamma_k$, $f^M(A) = 1$, and for all $B \in \Theta_k$, $f^M(B) = 0$. If $t = 1$, then, by (†), ($\P_1$), and ($\$$), $f^M(\oplus[A_0,\ldots, A_{n-1}]) = 0$. Similarly, if $t = 0$, then, by (†), ($\P_0$), and ($\$$), $f^M(\oplus[A_0,\ldots, A_{n-1}]) = 1$. Thus, in either case, $f^M(\oplus[A_0,\ldots, A_{n-1}]) \neq t$. On the other hand, since, by (†), $\{A_i : i \in I_k\} \subseteq \Gamma_k$ and $\{A_i : i \in J_k\} \subseteq \Theta_k$, this result is in contradiction with Lemma 3.3.1 (for, as was said before, L(M) is a common logic, and is certainly respected by **M**). Thus the *reductio* is complete.

Conversely, suppose that

($\$\$$) for all $k < q$, $\Gamma_k \vDash^M \Theta_k$.

Suppose then, for another *reductio ad absurdum*, that $\Gamma^* \nvDash^M \Theta^*$. Then there is a truth-value assignment f for the set of Boolean atoms such that for all $A \in \Gamma^*$, $f^M(A) = 1$, and for all $B \in \Theta^*$, $f^M(B) = 0$. In particular, by ($\P_1$) or ($\P_0$) – whichever is applicable – for all $A \in \Gamma$, $f^M(A) = 1$, for all $B \in \Theta$, $f^M(B) = 0$, and $f^M(\oplus[A_0,\ldots, A_{n-1}]) = t$. Let

$$I = \{i < n : f^M(A_i) = 1\}, \text{ and } J = \{i < n : f^M(A_i) = 0\}.$$

Then $\langle I, J \rangle$ is a partitioning of n; in fact, $\oplus$ is of type t with respect to it. Hence it follows from the hypothesis of the theorem that $I = I_k$ and $J = J_k$, for some $k < q$. But then we have an absurd situation: for all $A \in \Gamma_k, f^M(A) = 1$, and for all $B \in \Theta_k, f^M(B) = 0$ — absurd since it violates (\$\$). The *reductio* is complete. ∎

COROLLARY 3.7.2. $L_{\ddagger} \subseteq L(M)$.

Proof. By induction over $\langle B_{\ddagger}, R_{\ddagger} \rangle$. The basic step is obvious. The inductive step is taken care of by the lemma. ∎

The main effort in this section will be to show that, as far as finite sets go, $L_{\ddagger}$ and $L(M)$ agree. In other words, what will be shown is that, if $\mathbf{P}^{\circ}(\text{Form})$ denotes the set of finite sets of formulas, then

$$L_{\ddagger} \cap (\mathbf{P}^{\circ}(\text{Form}) \times \mathbf{P}^{\circ}(\text{Form})) = L(M) \cap (\mathbf{P}^{\circ}(\text{Form}) \times \mathbf{P}^{\circ}(\text{Form})).$$

In still other words — and taking the contents of Corollary 3.7.2 for granted —: M finitely respects $L_{\ddagger}$. This is the promised characterization of $L(M)$.

Central to our endeavour in this section will be the notion of search-tree. Before defining it, it is well to remind the reader that the material in Section 0.3, which is not required elsewhere in the book, is required here. As explained in that section, trees are functions whose domains are tree-structures, while the latter are a certain kind of sets of finite sequences of natural numbers. We shall employ the special conventions laid down in Section 0.3, according to which x, y, z are additional parameters for natural numbers, and **x, y, z** are parameters for sequences of natural numbers. If T is a tree, we shall use the notation dom T for the tree-structure of T (the domain of T). We shall sometimes attribute to a tree, without further explanation, properties that primarily belong to its underlying tree-structure. Thus we might say that a tree is finite if its tree-structure is; that **x** is a node of T if $\mathbf{x} \in \text{dom} T$; that Q is a branch through T if Q is a branch through dom T; etc.

We say that a function T is a *search-tree* if the following conditions are satisfied:

(i) T is a finitary tree;
(ii) The range of T is included in the set of pairs of formula sets;
(iii) For any $\langle x_0, \ldots, x_m \rangle$, suppose that $0, 1, \ldots, q-1$ are all the natural numbers k such that $\langle x_0, \ldots, x_m, k \rangle \in \text{dom} T$. Assume that $T(\langle x_0, \ldots, x_m \rangle) = \langle \Gamma^*, \Theta^* \rangle$ and that, for all $k < q$, $T(\langle x_0, \ldots, x_m, k \rangle) = \langle \Gamma_k, \Theta_k \rangle$. Then it is necessary that $\langle\langle \Gamma_0, \Theta_0 \rangle, \ldots, \langle \Gamma_{q-1}, \Theta_{q-1} \rangle, \langle \Gamma^*, \Theta^* \rangle\rangle \in R_{\ddagger}$;
(iv) For any $\langle x_0, \ldots, x_m \rangle$, if $T(\langle x_0, \ldots, x_m \rangle) = \langle \Gamma, \Theta \rangle$ and $\langle \Gamma, \Theta \rangle \in B_{\ddagger}$, then $\langle x_0, \ldots, x_m, 0 \rangle \notin \text{dom} T$.

We say that T is *finished* if the following additional condition is also satisfied:

(v) For every top node $x \in \text{dom}\,T$, if $T(x) = \langle \Gamma, \Theta \rangle$, then, unless $\langle \Gamma, \Theta \rangle \in B_\ddagger$, every element of Γ or Θ is either a Boolean atom or a Boolean propositional constant.

Finally, let us say that a finished search-tree T is a *proof-tree* if it is finite and, for every top node $z \in \text{dom}\,T$, if $T(z) = \langle \Gamma, \Theta \rangle$, then $\langle \Gamma, \Theta \rangle \in B_\ddagger$.

If T is a search-tree (proof-tree) such that $T(z) = \langle \Gamma, \Theta \rangle$, where z is the root, then we say that T is a search-tree (proof-tree) *for* $\langle \Gamma, \Theta \rangle$. Notice that there is no presumption of uniqueness here: except for degenerate cases, a pair of formula sets has several search-trees, and it has several proof-trees if it has one.

Before embarking on the next lemma, we pause to offer some informal remarks. The idea behind the introduction of search-trees is, roughly, to provide a decision procedure for what may be called the finite part of $L(M)$ — that is, the logic $L(M) \cap (P^\circ(\text{Form}) \times P^\circ(\text{Form}))$. Let us try to intimate what this means.

Consider how $L_\ddagger$ is generated. One starts out with the pairs of formula sets in $B_\ddagger$, then adds any pair that can be got by applying or keeping applying $R_\ddagger$. Whenever a pair has been obtained in this manner, it is possible to trace its derivational history (or, at least, to reconstruct a possible derivational history). In fact, the search-tree format provides a kind of normal form for the writing of this kind of history.

Thus, whenever one is interested in the question whether a certain pair of finite formula sets is an element of $L_\ddagger$, one may investigate whether there is a search-tree of the right kind. Placing $\langle \Gamma, \Theta \rangle$ at the root of the tree to be grown, one moves a step at a time. At each step one will ask, How might *this pair* have been obtained by an introduction rule $IAM(\oplus)$ or $ICM(\oplus)$? A positive answer to such a question allows the tree to grow by the addition of a number of new (immediately succeeding) nodes.

As one moves up through the tree, the participating formula sets get simpler and simpler (in a sense it remains to specify). Finally it comes to a point (a top node) where either one *need not* go further (the pair belongs to $B_\ddagger$) or else one *cannot* go further (the pair consists of two sets of Boolean atoms or Boolean propositional constants). At this stage, when the tree will grow no further, it is finished. According to the formal analysis to follow, the growth always ends in a finite number of steps. Moreover, the following will be seen to be the case. The root pair is an element of $L_\ddagger$ if and only if all the top pairs are elements of $B_\ddagger$. The root pair is an element of $L(M)$ if and only if all the top pairs are elements of $L(M)$. This allows the conclusion that the root pair is an element of $L_\ddagger$ if and only if it is an element of $L(M)$; this is the main result that we are after in this section, the remainder of which will lend substance to these remarks.

We adopt the following convention. If T is a search-tree which is defined for a certain sequence x, then we shall feel free to write $T(x) = \langle \Gamma_x, \Theta_x \rangle$.

BOOLEAN LOGICS 77

LEMMA 3.7.3. *Let T be a search-tree for a pair of finite formula sets. Then, for every* $x \in \text{dom} T$, $\langle \Gamma_x, \Theta_x \rangle$ *is a pair of finite sets.*

Proof. To a reader who has grasped the concept of search-tree, the lemma will seem self-evident. Nevertheless, let us give a rigorous proof.
  Suppose that T is a given search-tree with root z such that Γ_z and Θ_z are finite. We wish to show that, for every node x,

(£) $\qquad\qquad\qquad\Gamma_x$ and Θ_x are finite.

Let $\langle B_T, R_T \rangle$ be the upward inductive system associated with $\text{dom} T$ (see Section 0.3). The set of objects induced by this system coincides with $\text{dom} T$. We now prove our claim by induction over $\langle B_T, R_T \rangle$.
  Basic step. By assumption, (£) holds for the root of T.
  Inductive step. Let x by any node. Suppose, as our induction hypothesis, that (£) holds for x; that is, that Γ_x and Θ_x are finite. As T is a finitary tree, x has at most finitely many immediate successors $y_0, \ldots, y_{q-1}$. From the conditions on R_T, we infer that $x\ R_T\ y_0, \ldots, x\ R_T\ y_{q-1}$. By condition (iii) in the definition of search-tree,

$$\langle\langle \Gamma_{y_0}, \Theta_{y_0} \rangle, \ldots, \langle \Gamma_{y_{q-1}}, \Theta_{y_{q-1}} \rangle, \langle \Gamma_x, \Theta_x \rangle\rangle \in R_\ddagger.$$

But it is clear from the definition of $R_\ddagger$ that, since Γ_x and Θ_x are finite sets, then so are Γ_{y_k} and Θ_{y_k}, for all $k < q$. The induction is complete. ∎

If the preceding lemma seems trivial, the following one is, by comparison, a relatively deep result.

LEMMA 3.7.4. *Any search-tree for a pair of finite formula sets is finite.*

Proof. We need some concepts of a syntactic nature. The (*Boolean*) *degree* of a formula A, denoted by $\deg(A)$, is defined as follows:

$\deg(A) = 0$, if A is a Boolean atom;

$$= n + \sum_{i=0}^{n-1} \deg(A_i), \text{ if } \oplus \in \text{Bop and } n = r(\oplus), \text{ and } A = \oplus[A_0, \ldots, A_{n-1}].$$

The (*Boolean*) *profile* of a finite set Σ of formulas is the sequence $\langle x_i \rangle_{i<\omega}$ of natural numbers x_i such that, for each i, x_i is the number of formulas in Σ of degree i:

$$x_i = \text{card}\{A \in \Sigma: \deg(A) = i\},$$

where, for any set Δ, card Δ denotes the cardinality of Δ. It is clear that if $\langle x_i \rangle_{i<\omega}$ is the profile of a finite set, then there is some q such that, for all $i \geq q$, $x_i = 0$. Hence, the profiles of finite sets are virtually finite sequences of natural numbers, in the sense of Section 0.3.

After this preamble let us return to the proof proper. Suppose that T is a search-tree for some pair of finite sets. If T is not finite, then by König's Lemma (Theorem 0.3.7) there is some infinite branch Q through T. Let $\langle z_i \rangle_{i<\omega}$ be the sequence of nodes in Q such that, for each i, z_i is the immediate predecessor of z_{i+1} (this sequence exists, by Lemma 0.3.6). By Lemma 3.7.3, Γ_{z_i} and Θ_{z_i} are finite sets, for every i, so the profile of the set $\Gamma_{z_i} \cup \Theta_{z_i}$ is a well-defined quantity — let it be denoted by x_i.

Suppose now for the moment that the following claim is true:

(*) For every i, x_i dominates x_{i+1} in the sense of Section 0.3.

Then $\langle x_i \rangle_{i<\omega}$ is a dominance chain of virtually finite sequences of natural numbers, and indeed an infinite one. But this is in contradiction with Theorem 0.3.8. Thus the assumption that T contains an infinite branch is absurd if (*) is true. The rest of the proof is therefore devoted to a proof of (*).

Suppose, for any i, that $z_i = \langle z_0, \ldots, z_m \rangle$, and $z_{i+1} = \langle z_0, \ldots, z_m, j \rangle$. Let $0, 1, \ldots, q-1$ be all the natural numbers k such that $\langle z_0, \ldots, z_m, k \rangle \in \mathrm{dom}\, T$. Then, by condition (iii) in the definition of search-tree — and using the notation defined there — we conclude that

$$\langle\langle \Gamma_0, \Theta_0 \rangle, \ldots, \langle \Gamma_{q-1}, \Theta_{q-1} \rangle, \langle \Gamma^*, \Theta^* \rangle\rangle \in R_\ddagger.$$

Thus there is some $j < q$ such that

$$\langle \Gamma_j, \Theta_j \rangle = \langle \Gamma_{z_{i+1}}, \Theta_{z_{i+1}} \rangle,$$

while

$$\langle \Gamma^*, \Theta^* \rangle = \langle \Gamma_{z_i}, \Theta_{z_i} \rangle.$$

Going back to the definition of $R_\ddagger$, we conclude that there is an n-ary Boolean operator $\oplus$, for some n, some formulas $A_0, \ldots, A_{n-1}$, some formula sets Γ and Θ, some type $t \in \{0, 1\}$ such that (†) and (¶$_t$) hold, where $\langle I_0, J_0 \rangle, \ldots, \langle I_{q-1}, J_{q-1} \rangle$ are all the partitionings of n, and k ranges over the set $\{0, 1, \ldots, q-1\}$. It follows from condition (iv) in the definition of search-tree that $\langle \Gamma^*, \Theta^* \rangle \notin B_\ddagger$, and from this that $\Gamma^* \cap \Theta^* = \emptyset$. Therefore, if $t = 1$, then not only $\oplus[A_0, \ldots, A_{n-1}] \notin \Gamma$ but also $\oplus[A_0, \ldots, A_{n-1}] \notin \Theta$; and, similarly, if $t = 0$, then not only $\oplus[A_0, \ldots, A_{n-1}] \notin \Theta$ but also $\oplus[A_0, \ldots, A_{n-1}] \notin \Gamma$. Hence it is certain that $\oplus[A_0, \ldots, A_{n-1}] \notin \Gamma_j \cup \Theta_j$. This means that the difference between the pairs $\langle \Gamma^*, \Theta^* \rangle$ and $\langle \Gamma_j, \Theta_j \rangle$ can be expressed as follows:

(1) $\qquad (\Gamma^* \cup \Theta^*) - (\Gamma_j \cup \Theta_j) = \{\oplus[A_0, \ldots, A_{n-1}]\},$

(2) $\qquad (\Gamma_j \cup \Theta_j) - (\Gamma^* \cup \Theta^*) \subseteq \{A_i : i < n\}.$

Let us write $x_i = \langle x_{i,j} \rangle_{j<\omega}$, for all i. Suppose that $d = \deg(\oplus[A_0, \ldots, A_{n-1}])$. It follows from (1), (2), and the definition of profile that $x_{i,d} > x_{i+1,d}$ and that,

BOOLEAN LOGICS 79

for all $j > d$, $x_{i,j} = x_{i+1,j}$. Hence x_i dominates x_{i+1}, as we wanted to show. This concludes the proof of (*). ∎

THEOREM 3.7.5. *There exists a finished search-tree for every pair of finite sets of formulas.*

Proof. We introduce the concept of the (*Boolean*) *degree* of a finite set Σ of formulas, denoted by deg Σ:

$$\deg \Sigma = \sum_{i<\omega} d_i,$$

where, for each i, $d_i = \text{card}\{A \in \Sigma : \deg(A) = i\}$. (For the definition of deg(A), see the proof of the preceding lemma.) Since Σ is assumed to be finite, it is clear that deg Σ exists and is a natural number. Note that if both Σ and Δ are finite sets, then always $\deg(\Sigma \cup \Delta) \leq \deg\Sigma + \deg\Delta$, and that $\Sigma \cap \Delta = \emptyset$ implies $\deg(\Sigma \cup \Delta) = \deg\Sigma + \deg\Delta$.

We prove the theorem by course-of-values induction. Thus let j be a fixed natural number and assume, as the induction hypothesis, that

(§)   if Γ and Θ are finite sets of formulas such that $\deg(\Gamma \cup \Theta) < j$, then there is a finished search-tree for $\langle \Gamma, \Theta \rangle$.

Let Γ^* and Θ^* be particular finite sets of formulas such that

(1) $\deg(\Gamma^* \cup \Theta^*) = j$.

It will now be enough to establish the existence of a finished search-tree for $\langle \Gamma^*, \Theta^* \rangle$.

Case 1: $\langle \Gamma^*, \Theta^* \rangle \in B_{\ddagger}$. In this case, $\langle\langle 0 \rangle, \langle \Gamma^*, \Theta^* \rangle\rangle$ is a finished search-tree.
Case 2: $\langle \Gamma^*, \Theta^* \rangle \in B_{\ddagger}$. There are two subcases.

Case 2.1: No formula of Γ^* or Θ^* contains a Boolean operator of rank >0. In this case, Γ^* and Θ^* are sets of Boolean atoms and Boolean propositional constants. It will be enough to establish the following claim for all p:

(£)   If the number of Boolean propositional constants in $\Gamma^* \cup \Theta^*$ is p, then there is a finished search-tree for $\langle \Gamma^*, \Theta^* \rangle$.

This we propose to do by ordinary induction on p.
Basic step. Suppose that Γ^* and Θ^* are sets of Boolean atoms. Then $\langle\langle 0 \rangle, \langle \Gamma^*, \Theta^* \rangle\rangle$ is a finished search-tree for $\langle \Gamma^*, \Theta^* \rangle$. This shows that (£) holds for p = 0.
Inductive step. Suppose, as the induction hypothesis, that (£) holds for some fixed number p. Suppose that $\Gamma^* \cup \Theta^*$ contains p + 1 Boolean propositional constants; let $\triangle$ be one of them. Then $\mathsf{M}(\triangle)$ is a constant truth-value function, either T or F. Note that $\mathsf{M}(\triangle) \equiv 1$ if and only if $\mathsf{M}(\triangle) = T$, and that $\mathsf{M}(\triangle) \equiv 0$ if and only if $\mathsf{M}(\triangle) = F$.

First suppose that $\mathsf{M}(\triangle) = T$. If $\triangle \in \Theta^*$ we would have to conclude that

$\langle \Gamma^*, \Theta^* \rangle \in B_1$, which is impossible under Case 2. Hence $\triangle \notin \Theta^*$; and so $\triangle \in \Gamma^*$. Let us write $\Gamma = \Gamma^* - \{\triangle\}$, and $\Theta = \Theta^*$. It follows from the definition of $R_\ddagger$ that $\langle\langle \Gamma, \Theta \rangle, \langle \Gamma^*, \Theta^* \rangle\rangle \in R_\ddagger$. Now, $\Gamma \cup \Theta$ contains exactly p Boolean propositional constants. Hence, by the induction hypothesis, there is a finished search-tree T for $\langle \Gamma, \Theta \rangle$. We define a new tree T^* as follows. First the domain of T^*:

$$\text{dom}\,T^* = \{\langle 0 \rangle\} \cup \{\langle 0, x_0, \ldots, x_{m-1} \rangle : \langle x_0, \ldots, x_{m-1} \rangle \in \text{dom}\,T\}.$$

Then the tree itself:

$$T^*(\langle 0 \rangle) = \langle \Gamma^*, \Theta^* \rangle,$$

$$T^*(\langle 0, x_0, \ldots, x_m \rangle) = T(\langle x_0, \ldots, x_{m-1} \rangle).$$

It is clear that T^* is a finished search-tree for $\langle \Gamma^*, \Theta^* \rangle$.

If instead $\mathbf{M}(\triangle) = F$, we can proceed in an entirely analogous manner. Therefore the ordinary induction is complete.

Case 2.2: There is some formula $\oplus[A_0, \ldots, A_{n-1}] \in \Gamma^* \cup \Theta^*$, where $\oplus$ is an n-ary Boolean operator and

(2) $\qquad\qquad\qquad\qquad n > 0.$

Define a type $t \in \{0, 1\}$ by the requirement that

$$t = 1, \text{ if } \oplus[A_0, \ldots, A_{n-1}] \in \Gamma^*,$$
$$= 0, \text{ if } \oplus[A_0, \ldots, A_{n-1}] \in \Theta^*.$$

Under Case 2, $\Gamma^* \cap \Theta^* = \emptyset$, so t is well-defined. Let $\langle I_0, J_0 \rangle, \ldots, \langle I_{q-1}, J_{q-1} \rangle$ be all the partitionings of n with respect to which $\oplus$ is of type t in $L(\mathbf{M})$. Notice that $q > 0$. For suppose $q = 0$. Then there is no partitioning of n with respect to which $\oplus$ is of type t. Accordingly $\mathbf{M}(\oplus)$ must be a constant function which is identically $1 - t$. Thus if $t = 0$, then $\mathbf{M}(\oplus) \equiv 1$, and so $\langle \Gamma^*, \Theta^* \rangle \in B_1$, which is impossible under Case 2. Similarly, if $t = 0$, then $\mathbf{M}(\oplus) \equiv 0$, and so $\langle \Gamma^*, \Theta^* \rangle \in B_0$, which is similarly impossible.

Let Γ and Θ be formula sets such that condition ($\P_t$) in the definition of $R_\ddagger$ is satisfied. For each $k < q$, adopt condition (†) of that same definition as a definition here. It follows that

$$\langle\langle \Gamma_0, \Theta_0 \rangle, \ldots, \langle \Gamma_{q-1}, \Theta_{q-1} \rangle, \langle \Gamma^*, \Theta^* \rangle\rangle \in R_\ddagger.$$

By (†), for each $k < q$, $\Gamma_k \cup \Theta_k = (\Gamma \cup \Theta) \cup \{A_i : i < n\}$. Hence,

(3) $\qquad \deg(\Gamma_k \cup \Theta_k) \leqq \deg(\Gamma \cup \Theta) + \sum_{i<n} \deg(A_i), \text{ for each } k < q.$

If $t = 1$, it follows from ($\P_1$) that $\oplus[A_0, \ldots, A_{n-1}] \in \Gamma^* - \Gamma$ and that $\Theta^* = \Theta$; since, under Case 2, $\Gamma^* \cap \Theta^* = \emptyset$, therefore, $\oplus[A_0, \ldots, A_{n-1}] \notin \Theta$. If $t = 0$, it follows from (¶$_0$) that we instead have $\oplus[A_0,\ldots,A_{n-1}] \in \Theta^* - \Theta$, and that $\Gamma^* = \Gamma$; and so, for the same reason, $\oplus[A_0,\ldots,A_{n-1}] \notin \Gamma$. Thus, in either case, $\oplus[A_0,\ldots,A_{n-1}] \notin \Gamma \cup \Theta$. Therefore,

(4) $\qquad \deg(\Gamma^* \cup \Theta^*) = \deg(\Gamma \cup \Theta) + \deg(\oplus[A_0,\ldots,A_{n-1}])$.

But, by the definition of degree (see the proof of Lemma 3.7.4),

(5) $\qquad \deg(\oplus[A_0,\ldots,A_{n-1}]) = n + \sum_{i<n} \deg(A_i)$.

(1)-(5) imply that $\deg(\Gamma_k \cup \Theta_k) < j$, for each $k < q$. This means that (§) — this is the first time that this induction hypothesis comes into play! — applies to each $\langle \Gamma_k, \Theta_k \rangle$, yielding finished search-trees T_0 for $\langle \Gamma_0, \Theta_0 \rangle, \ldots, T_{q-1}$ for $\langle \Gamma_{q-1}, \Theta_{q-1} \rangle$. Since, as we saw, $q > 0$, this collection of trees is non-empty.

We now construct a new tree T^* as follows. First we define $\mathrm{dom}\, T^*$, the domain of the tree:

$$\mathrm{dom}\, T^* = \{\langle 0 \rangle\} \cup \{\langle 0, k, x_1, \ldots, x_{m-1}\rangle : k < q \,\&\, \langle x_0, x_1, \ldots, x_{m-1}\rangle \in \mathrm{dom}\, T_k\}.$$

Evidently $\mathrm{dom}\, T^*$ is a tree-structure. Now T^* itself is defined on its intended domain as follows:

$$T^*(\langle 0 \rangle) = \langle \Gamma^*, \Theta^* \rangle.$$

$$T^*(\langle 0, k, x_1, \ldots, x_{m-1}\rangle) = T_k(\langle x_0, x_1, \ldots, x_{m-1}\rangle).$$

It is easy to verify that T^* is a finished search-tree for $\langle \Gamma^*, \Theta^* \rangle$. The course-of-values induction is complete. ∎

THEOREM 3.7.6. *Let T be a finished search-tree for some pair $\langle \Gamma, \Theta \rangle$ of finite formula sets. Then $\Gamma \vDash^M \Theta$ only if T is a proof-tree.*

Proof. Suppose that T is a finished proof-tree for $\langle \Gamma, \Theta \rangle$ where Γ and Θ are finite sets of formulas. Assume that $\Gamma \vDash^M \Theta$. We shall begin by proving the claim that, for all nodes x of T, $\Gamma_x \vDash^M \Theta_x$. By Theorem 0.3.2, this may be proved by induction over the upward system associated with $\mathrm{dom}\, T$; call this system $\langle B', R' \rangle$.

Basic step. Suppose that $x \in B'$. Then x is the root of T. Since T is a search-tree for $\langle \Gamma, \Theta \rangle$, it follows that $\Gamma = \Gamma_x$ and $\Theta = \Theta_x$. Hence, by assumption, $\Gamma_x \vDash^M \Theta_x$.

Inductive step. Suppose that x and y are fixed nodes such that $x\, R'\, y$. Assume, as the inductive hypothesis, that

(§) $\qquad\qquad\qquad \Gamma_x \vDash^M \Theta_x.$

Since T is a finitary tree, x has at most finitely many immediate successors, say $y_0, \ldots, y_{q-1}$; evidently, y is one among them. It follows that

$$\langle\langle\Gamma_{y_0}, \Theta_{y_0}\rangle, \ldots, \langle\Gamma_{y_{q-1}}, \Theta_{y_{q-1}}\rangle, \langle\Gamma_x, \Theta_x\rangle\rangle \in R_{\ddagger}.$$

Hence, by (§) and Lemma 3.7.1, $\Gamma_{y_k} \vDash^M \Theta_{y_k}$, for all $k < q$. In particular, $\Gamma_y \vDash^M \Theta_y$. The induction is complete and the claim proved.

Now since by assumption T is a finished search-tree, it is finite (Lemma 3.7.4). Let z be any top node; then Γ_z and Θ_z are sets of Boolean atoms and Boolean propositional constants. We must show that $\langle\Gamma_z, \Theta_z\rangle \in B_{\ddagger}$. Suppose first that there is some Boolean propositional constant $\triangle \in \Gamma_z \cup \Theta_z$. There are two cases. The first case is when $M(\triangle) = T$. If $\triangle \in \Gamma_z$, then T would not be finished; for it is clear that $\langle\langle\Gamma_z - \{\triangle\}, \Theta_z\rangle, \langle\Gamma_z, \Theta_z\rangle\rangle \in R_{\ddagger}$. Hence $\triangle \in \Theta_z$ and so $\langle\Gamma_z, \Theta_z\rangle \in B_1$. The second case is when $M(\triangle) = F$. An analogous argument then yields $\langle\Gamma_z, \Theta_z\rangle \in B_0$.

Next suppose that $\Gamma_z \cup \Theta_z$ contains no Boolean propositional constant; hence Γ_z and Θ_z are now sets of Boolean atoms. Let v be any truth-value assignment such that, for every Boolean atom A,

$$v(A) = 1, \text{ if } A \in \Gamma_z - \Theta_z,$$
$$= 0, \text{ if } A \in \Theta_z - \Gamma_z.$$

By the claim established in the first part of this proof, $\Gamma_z \vDash^M \Theta_z$. Hence, either there is some $A \in \Gamma_z$ such that $v^M(A) = 0$ and therefore $A \notin \Gamma_z - \Theta_z$, or there is some $B \in \Theta_z$ such that $v^M(B) = 1$ and therefore $B \notin \Theta_z - \Gamma_z$. In either case, $\Gamma_z \cap \Theta_z \neq \emptyset$, and so $\langle\Gamma_z, \Theta_z\rangle \in B_{\text{diag}}$. ∎

THEOREM 3.7.7. *Let T be a proof-tree for some pair $\langle\Gamma, \Theta\rangle$ of finite formula sets. Then $\langle\Gamma, \Theta\rangle \in L_{\ddagger}$.*

Proof. Assume that T is a proof-tree for $\langle\Gamma, \Theta\rangle$, where Γ and Θ are finite sets of formulas. It will be enough to establish the claim that, for every node x of T, $\langle\Gamma_x, \Theta_x\rangle \in L_{\ddagger}$. By Theorem 0.3.3, this may be proved by induction over the downward inductive system associated with dom$T$; let us call this system $\langle\overline{B}, \overline{T}\rangle$.

Basic step. If $x \in \overline{B}$, then x is a top. Since T is a proof-tree, this means that $\langle\Gamma_x, \Theta_x\rangle \in B_{\ddagger}$, and so, *a fortiori*, $\langle\Gamma_x, \Theta_x\rangle \in L_{\ddagger}$.

Inductive step. Fix any nodes $x, y_0, \ldots, y_{q-1}$ such that

(1) $\qquad\qquad\qquad \langle y_0, \ldots, y_{q-1}, x\rangle \in \overline{R}.$

Assume, as our induction hypothesis, that

(§) $\qquad\qquad\qquad \langle\Gamma_{y_k}, \Theta_{y_k}\rangle \in L_{\ddagger}$, for all $k < q$.

From (1) it follows that $y_0, \ldots, y_{q-1}$ are all the immediate successors of x in T. Hence, by condition (iii) in the definition of search-tree,

(2) $\qquad \langle\langle\Gamma_{y_0}, \Theta_{y_0}\rangle, \ldots, \langle\Gamma_{y_{q-1}}, \Theta_{y_{q-1}}\rangle, \langle\Gamma_x, \Theta_x\rangle\rangle \in R_{\ddagger}.$

From (2) and (§) it follows at once that $\langle\Gamma_x, \Theta_x\rangle \in L_{\ddagger}$. This completes both the induction and the proof. ∎

BOOLEAN LOGICS   83

With all the preceding preparations, the desired main result is now an easy corollary:

THEOREM 3.7.8. *Let Γ and Θ be any finite sets of formulas. Then $\langle \Gamma, \Theta \rangle \in L_{\ddagger}$ if and only if $\Gamma \vDash^M \Theta$.*

Proof. One half of the theorem has already been proved (Corollary 3.7.2). For the remaining half — the 'completeness part' — assume that $\Gamma \vDash^M \Theta$, for some finite sets Γ and Θ of formulas. By Theorem 3.7.5, there is a finished search-tree for $\langle \Gamma, \Theta \rangle$. By Theorem 3.7.6, this tree is a proof-tree. Hence, by Theorem 3.7.7, $\langle \Gamma, \Theta \rangle \in L_{\ddagger}$. ∎

Some final remarks. In our terminology, $\langle B_{\ddagger}, R_{\ddagger} \rangle$ is an inductive system. In more orthodox parlance, it is an *axiom system*: the elements of $B_{\ddagger}$ are the *axioms*, and the elements of $R_{\ddagger}$ can be subsumed under certain *inference rules*, viz., (**IAM**($\oplus$)) and (**ICM**($\oplus$)), for all Boolean operators $\oplus$. It is of interest to note that no elimination rules and no monotonicity or cut rules were assumed in the definition of $L_{\ddagger}$. To be sure, the elimination rules (**EAM**($\oplus$)) and (**ECM**($\oplus$)) hold in $L_{\ddagger} \cap (\mathbf{P}°(\text{Form}) \times \mathbf{P}°(\text{Form}))$, which is also closed under both (Mono) and (Cut). These rules may therefore be labelled *derived rules*, but *primitive* in $\langle B_{\ddagger}, R_{\ddagger} \rangle$ they are not. Results of this sort — especially Cut Elimination Theorems or *Hauptsätze* — play a great role in ordinary Gentzen theory.

We said above that the notion of search-tree was introduced in order to develop a decision procedure. Now we can see how it works. Given any pair $\langle \Gamma, \Theta \rangle$ of finite sets of formulas, we can construct, in a finite number of steps, a finished search-tree T for this pair. It may then be checked, again in a finite number of steps, whether T is a proof-tree. If it is, then we have what would normally be called a *proof* of $\langle \Gamma, \Theta \rangle$. If it is not, then we implicitly have what can be called a *disproof* of $\langle \Gamma, \Theta \rangle$; for we may extract from T a valuation which assigns 1 to the elements of Γ and 0 to the elements of Θ.

The last of the preceding observations is worth an elaboration. Suppose that T fails to be a proof-tree. Then there must be some top node x such that $\langle \Gamma_x, \Theta_x \rangle \notin B_{\ddagger}$, where Γ_x and Θ_x are finite sets of Boolean atoms and Boolean propositional constants. Let v be any truth-value assignment for the set of Boolean atoms such that, for every Boolean atom A,

$$v(A) = 1, \text{ if } A \in \Gamma_x,$$
$$= 0, \text{ if } A \in \Theta_x.$$

That such a truth-value assignment exists must of course be argued for. In fact, it owes its existence to the assumption that $\langle \Gamma_x, \Theta_x \rangle \notin B_{\ddagger}$. First, since $\langle \Gamma_x, \Theta_x \rangle \notin B_{\text{diag}}$, and so $\Gamma_x \cap \Theta_x = \emptyset$, as far as Boolean atoms go, the condition on v is unproblematic. Second, since $\langle \Gamma_x, \Theta_x \rangle \notin B_0$, there is no Boolean propositional constant $\triangle \in \Gamma_x$ such that $M(\triangle) = F$; therefore, if $\triangle \in \Gamma_x$, then $M(\triangle) = T$. In other words, no Boolean propositional constant in Γ_x can violate the condition on v. Third, since $\langle \Gamma_x, \Theta_x \rangle \notin B_1$, it follows by an analogous argument that no Boolean propositional constant in Θ_x can violate that condition either. Consequently, v can be found. (Notice that all this holds even if Γ_x, or Θ_x, or even both are empty.)

Since x is a top node, it determines a branch Q through T. Hence, by Lemma 0.3.6, there are natural numbers $x_0, \ldots, x_m$ such that $Q = \{x_i : i \leq m\}$, where, for each $i \leq m$, $x_i = \langle x_0, \ldots, x_i \rangle$. Now we claim that, for each natural number $i \leq m$, it holds for every formula A — not necessarily a Boolean atom or Boolean propositional constant — that

$$v^M(A) = 1, \text{ if } A \in \Gamma_{x_i},$$
$$= 0, \text{ if } A \in \Theta_{x_i}.$$

This claim can be proved by backward induction from m (cf. the proof of Lemma 0.3.1). By now the readers will have had their fill of such proofs, so perhaps it is as well to omit the details. However, consider the particular instance $i = 0$: the claim — or, when proved, the result — that, for all formulas A,

$$v^M(A) = 1, \text{ if } A \in \Gamma_{x_0},$$
$$= 0, \text{ if } A \in \Theta_{x_0}.$$

Since x_0 is the root of T, this is what we want: it shows that $\Gamma \not\Vdash^M \Theta$ (and hence, by Corollary 3.7.2, that $\langle \Gamma, \Theta \rangle \notin L_\ddagger$).

4
PRE-CLASSICAL LOGICS

4.1. Expressibility

The next step on our long path to a definition of classical logics is the introduction of a family of logics that we will call 'pre-classical'. They are the logics in which, roughly speaking, all truth-value functions are 'expressible'; but this notion we have not yet defined, and this we must now do.

Let some fixed language be given. Let M be a matrix for it. Let ϕ be an n-ary truth-value function. Let A be a purely Boolean formula and $P_0, \ldots, P_{n-1}$ distinct propositional letters. Then we say that A *expresses ϕ under* M *with respect to* $P_0, \ldots, P_{n-1}$ if, for all truth-value assignments f for the set of propositional letters in A we have

$$\bar{f}A = \phi(fP_0, \ldots, fP_{n-1}).$$

Here the order of the propositional letters is important. As a function, ϕ maps n-tuples of truth-values to truth-values. Thus, for each $i < n$, P_i corresponds to the i'th coordinate of the n-tuples.

We say that an n-ary truth-value function ϕ is *expressible under* M if there is some formula A expressing it under M with respect to some $P_0, \ldots, P_{n-1}$. This terminology is in harmony with the one laid down in the preceding section: If an n-ary operator ✩ directly expresses an n-place truth-value function ϕ under M, then ϕ is expressible under M — for any distinct propositional letters $P_0, \ldots, P_{n-1}$, the formula ✩$[P_0, \ldots, P_{n-1}]$ expresses ϕ under M with respect to $P_0, \ldots, P_{n-1}$. If L is a consistent Boolean logic and M is the matrix implicit in L, then we may use the preceding locutions with 'under M' replaced by 'in L'. Reference to M or L is often suppressed when clarity is not thereby endangered.

LEMMA 4.1.1. *Let* L *be any consistent Boolean logic and ϕ an n-place truth-value function. If both* A *and* A' *express ϕ in* L *with respect to* $P_0, \ldots, P_{n-1}$, *then* A ⊣⊢ A'.

Proof. Let f be any truth-value assignment for $\{P_0, \ldots, P_{n-1}\}$, then $\bar{f}A = \phi(fP_0, \ldots, fP_{n-1}) = \bar{f}A'$. Hence A ⊣⊨ A', and so, by respect, A ⊣⊢ A'. ∎

The remaining results in this section, including Lemma 4.1.3, may seem obvious (if the reader is able to see through the technical paraphernalia to the contents). As sometimes happens with 'obvious' results, it is a bit cumbersome to give rigorous proofs for them. It depends on the reader's taste whether such proofs seem called for. It is only fair to include reasonably careful proofs here; but readers who want to skip them may do so, knowing that nothing in those proofs is a prerequisite for the understanding of later parts of the book.

LEMMA 4.1.2. *Let L be a consistent Boolean logic, and let A be any purely Boolean formula with propositional letters* $P_0, \ldots, P_{n-1}$. *Then there is an n-ary truth-value function* ϕ *such that A expresses* ϕ *with respect to* $P_0, \ldots, P_{n-1}$.

Proof. This result is a corollary of Theorem 3.3.3. ■

This lemma is presupposed in the statement of the following rather technical lemma.

LEMMA 4.1.3. *Let L be a consistent Boolean logic. Let* $\phi, \psi_0, \ldots, \psi_{n-1}$ *be truth-value functions of arity* n, $k_0, \ldots, k_{n-1}$, *respectively. Suppose that* A, $B_0, \ldots, B_{n-1}$ *are purely Boolean formulas such that A expresses* ϕ *with respect to* $P_0, \ldots, P_{n-1}$, *and, for all* $i < n$, B_i *expresses* ψ_i *with respect to* $Q_0^i, \ldots, Q_{k_i-1}^i$. *Let s be the substitution function such that* $sP_i = B_i$, *for each* $i < n$, *and s agrees with the identity function elsewhere. Let* $Q_0, \ldots, Q_{p-1}$ *be all distinct propositional letters in the set* $\{Q_j^i : i < n \ \& \ j < k_i\}$. *Let* χ *be the p-ary truth-value function expressed by sA with respect to* $Q_0, \ldots, Q_{p-1}$. *Then* $\chi = \phi(\psi_0, \ldots, \psi_{n-1})$.

Proof. It would be enough to prove that if f is any truth-value assignment for the set of propositional letters in sA, then

(‡) $\quad \bar{f}(sA) = \phi(\psi_0(fQ_0^0, \ldots, fQ_{k_0-1}^0), \ldots, \psi_{n-1}(fQ_0^{n-1}, \ldots, fQ_{k_{n-1}-1}^{n-1}))$.

Since, for each $i < n$, f is a truth-value assignment for the set of propositional letters in B_i, we know that

(1) $\qquad\qquad \bar{f}B_i = \psi_i(fQ_0^i, \ldots, fQ_{k_i-1}^i)$.

Moreover, if g is any truth-value assignment for the set of propositional letters in A, we know that

(2) $\qquad\qquad \bar{g}A = \phi(gP_0, \ldots, gP_{n-1})$.

Let h be any truth-value assignment for the set of propositional letters in A such that, for all $i < n$,

(3) $\qquad\qquad hP_i = \bar{f}B_i$.

We claim that, for all Boolean combinations C of propositional letters in A,

(4) $$\bar{h}C = \bar{f}(sC).$$

The claim is to be proved by induction on C.

If C is a propositional letter, the claim is true by definition of h. If C is not a propositional letter, then, since C is purely Boolean, there must be some Boolean operator ☆, q-ary say, such that $C = ☆[D_0,\ldots, D_{q-1}]$, for some $D_0,\ldots, D_{q-1}$. As our inductive hypothesis assume that the claim holds for $D_0,\ldots, D_{q-1}$. Let **M** be the matrix that is implicit in L. Then we have

$$h(☆[D_0,\ldots, D_{q-1}]) = \mathbf{M}(☆)(\bar{h}D_0,\ldots, \bar{h}D_{q-1})$$
$$= \mathbf{M}(☆)(\bar{f}(sD_0),\ldots, \bar{f}(sD_{q-1}))$$
$$= \bar{f}(☆[sD_0,\ldots, sD_{q-1}])$$
$$= \bar{f}s(☆[D_0,\ldots, D_{q-1}]).$$

Thus (4) holds.

Appealing to the fact that, by assumption, A is a Boolean combination, and using (4), (2), (3), and (1), in that order, we conclude that

$$\bar{f}(sA) = \bar{h}A$$
$$= \phi(hP_0,\ldots, hP_{n-1})$$
$$= \phi(\bar{f}B_0,\ldots, \bar{f}B_{n-1})$$
$$= \phi(\psi_0(fQ_0^0,\ldots, fQ_{k_0-1}^0),\ldots, \psi_{n-1}(fQ_0^{n-1},\ldots, fQ_{k_{n-1}-1}^{n-1})).$$

But this is just the condition (‡) we want. ∎

COROLLARY 4.1.4. *Let L be a consistent Boolean logic. Suppose that* A *expresses an n-ary truth-value function* ϕ *with respect to* $P_0,\ldots, P_{n-1}$. *Let* $Q_0,\ldots, Q_{n-1}$ *be any distinct propositional letters not occurring in* A. *Then* $A(P_0/Q_0,\ldots, P_{n-1}/Q_{n-1})$ *expresses* ϕ *with respect to* $Q_0,\ldots, Q_{n-1}$.

Proof. The well-known lambda notation would be useful here. But since we have not introduced it, let us offer the following considerations instead.

The formula A expresses ϕ with respect to $P_0,\ldots, P_{n-1}$. The n-ary truth-value function ϕ maps n-tuples of truth-values to truth-values. We might emphasize this in our notation by writing

$$\phi = \phi(\underbrace{\cdot,\ldots, \cdot}_{n})$$

On the other hand, for each $i < n$, Q_i expresses the identity function with respect to Q_i. The identity function is a unary truth-value function, mapping (1-tuples of) truth-values to truth-values. Writing ι for this function, we may represent it as

$$\iota = \iota(\cdot).$$

By Lemma 4.1.3, then, $A(P_0/Q_0,\ldots, P_{n-1}/Q_{n-1})$ expresses

$$\underbrace{\phi(\iota(\cdot),\ldots, \iota(\cdot))}_{n}$$

with respect to $Q_0,\ldots, Q_{n-1}$. But $\phi(\iota(\cdot),\ldots, \iota(\cdot)) = \phi$. ∎

THEOREM 4.1.5. *In a consistent Boolean logic, the following conditions are equivalent:*

(i) *Every truth-value function is expressible;*
(ii) *F and C are expressible;*
(iii) *N and C are expressible;*
(iv) *N and K are expressible;*
(v) *N and A are expressible.*

Proof. This is a consequence of Lemma 4.1.3 and some remarks in section 3.6. ∎

4.2. Boolean extensions

Let us consider the following familiar situation. Suppose that L is a Boolean logic. Suppose furthermore that we are interested in a certain n-place truth-value function ϕ that is expressible in L, but not directly so. If we wanted to make ϕ directly expressible, then a very natural (and often used) procedure would be the following. Take any symbol ☆ that is not a primitive symbol of the language, and let ☆ be added to our stock of Boolean operators with the proviso that its rank is to be n. Since ϕ is expressible in L there will be some formula A with the following properties: (i) A is a Boolean compound of propositional letters among which are some n distinct propositional letters $P_0,\ldots, P_{n-1}$; (ii) A expresses ϕ in L with respect to $P_0,\ldots, P_{n-1}$. Let L* be the smallest common logic (in the augmented language) extending L in which A and ☆$[P_0,\ldots, P_{n-1}]$ are interdeducible. Then, with a vague formulation, one would expect that L* is nothing but L with ϕ made directly expressible. That such an expectation is well founded will be argued presently. But first we must provide a formal framework in which the informally described situation can be reviewed in exact terms.

Let $\mathcal{L}_1 = \langle \text{Lett}_1, \text{Bop}_1, \text{Iop}_1, r_1 \rangle$ and $\mathcal{L}_0 = \langle \text{Lett}_0, \text{Bop}_0, \text{Iop}_0, r_0 \rangle$ be propositional languages and ☆ a symbol such that

(i) $\quad\quad\quad\quad\quad\quad \text{Lett}_1 = \text{Lett}_0;$
(ii) $\quad\quad\quad\quad\quad\quad \text{Bop}_1 = \text{Bop}_0 \cup \{☆\};$
(iii) $\quad\quad\quad\quad\quad\quad \text{Iop}_1 = \text{Iop}_0;$

(iv) r_1 and r_0 agree on $Bop_0 \cup Iop_0$.

Let L_1 be a logic in $\mathcal{L}_1$ and L_0 one in $\mathcal{L}_0$. Suppose that A is a purely Boolean formula in $\mathcal{L}_0$ containing at least the propositional letters $P_0, \ldots, P_{n-1}$ (it may contain others as well). Finally, suppose that L_1 is the intersection of all common logics L in $\mathcal{L}_1$ such that

(v) $\qquad\qquad\qquad L \supseteq L_0$,

(vi) for all expressions E, F and formulas $C_0, \ldots, C_{n-1}$, if $E \star \star [C_0, \ldots, C_{n-1}] \star F$ is a formula of $\mathcal{L}_1$, then

$$E \star \star [C_0, \ldots, C_{n-1}] \star F \dashv\vdash_L E \star A(C_0, \ldots, C_{n-1}) \star F.$$

Then, by the definition in Section 2.5, L_1 is a definitional extension of L_0, one that we will call an *immediate Boolean extension* of L_0. The condition

$$\star [P_0, \ldots, P_{n-1}] \dashv\vdash_{L_1} A$$

will be referred to as *the definition of $\star$ in* L_1, and A as *the defining formula* of the definition.

The concept 'Boolean extension of' is defined as the reflexive, transitive closure of the concept 'immediate Boolean extension of'. That is to say, L* is a *Boolean extension of* L if and only if there are logics $L_0, \ldots, L_n$ such that $L_0 = L$, $L_n = L^*$, and, for all $i < n$, L_{i+1} is an immediate Boolean extension of L_i. (Warning: this terminology is a bit insidious in that a logic may be Boolean, extending another logic, without being a Boolean extension of it.)

THEOREM 4.2.1. *Let L be a Boolean logic and L* a Boolean extension of L. Then*

(i) *L* is a Boolean logic;*
(ii) *L* is conservative over L;*
(iii) *L* is definitional over L;*
(iv) *L and L* are syntactically equivalent.*

Proof. It will be enough to prove the theorem on the assumption that L* is an immediate Boolean extension of L.

Let Form and Form* be the respective sets of formulas in the languages of L and L*. Let $\star$ be the unique Boolean operator, n-ary say, in the language of L* that is not also in the language of L. Let $A \in$ Form be a Boolean compound of some distinct propositional letters $P_0, \ldots, P_{n-1}$ such that L* is the smallest common logic satisfying the two conditions

(1) $\qquad\qquad\qquad L^* \supseteq L$,

(2) $\qquad E \star \star [C_0, \ldots, C_{n-1}] \star F \dashv\vdash_{L^*} E \star A(C_0, \ldots, C_{n-1}) \star F$,

for all expressions E, F and formulas $C_0, \ldots, C_{n-1}$ such that $E \star \star [C_0, \ldots, C_{n-1}] \star F$ is a formula in the language of L*.

Part (i). First we wish to prove that L* is Boolean. By virtue of Theorem 3.4.3 and the fact that L* extends L, it will be enough to show that ☆ is type determined. Let $\langle I, J \rangle$ be any partitioning of n. By Theorem 3.3.3 and (1), we have

$$A, \{P_i : i \in I\} \vdash_{L^*} \{P_i : i \in I\}, \text{ or } \{P_i : i \in I\} \vdash_{L^*} \{P_i : i \in J\}, A.$$

If in (2) we take E and F as the empty expression and let $C_0, \ldots, C_{n-1}$ be $P_0, \ldots, P_{n-1}$, respectively, then we see that $\star[P_0, \ldots, P_{n-1}] \dashv\vdash_{L^*} A$. By definition, L* is common and therefore closed under cut. Hence either

$$\star[P_0, \ldots, P_{n-1}], \{P_i : i \in I\} \vdash_{L^*} \{P_i : i \in J\},$$

or else

$$\{P_i : i \in I\} \vdash_{L^*} \{P_i : i \in J\}, \star[P_0, \ldots, P_{n-1}].$$

Therefore ☆ is type determined in L* (cf. remarks in section 3.3).

Part (ii). Next we wish to prove that L* is conservative over L. Consider the following recursive definition of a function ° from Form* to Form:

$$P° = P, \text{ if P is a propositional letter};$$
$$\oplus[B_0, \ldots, B_{m-1}]° = \oplus[B_0°, \ldots, B_{m-1}°], \text{ if } \oplus \neq \star;$$
$$= A(P_0/B_0°, \ldots, P_{n-1}/B_{n-1}°), \text{ if } \oplus = \star \text{ (and hence m = n)}.$$

We write $\Sigma° = \{B° : B \in \Sigma\}$. Hence $\Sigma° \subseteq$ Form, if $\Sigma \subseteq$ Form*. Define

$$L^\# = \{\langle \Gamma, \Theta \rangle : \langle \Gamma°, \Theta° \rangle \in L\}.$$

We claim that $L^* \subseteq L^\#$.

To prove this claim it will be enough to show that $L^\#$ is a common logic (for clearly $L^\#$ includes L, and L* is the smallest common logic to do so). That $L^\#$ is indeed a common logic — that the four conditions (Refl), (Mono), (Cut), and (Subst) are all satisfied — is easy to see. As an example we give the argument for (Subst).

First note that if s is any substitution function, then, for every formula $B \in$ Form*, $s(B°) = (sB)°$. This can be proved by a straightforward inductive argument. Hence also for every set $\Sigma \subseteq$ Form*, $s(\Sigma°) = (s\Sigma)°$, where as usual $s\Sigma = \{sB : B \in \Sigma\}$ (and of course $s(\Sigma°) = \{sB : B \in \Sigma°\}$).

Suppose now that $\langle \Gamma, \Theta \rangle \in L^\#$ and let s be any substitution function in $\mathcal{L}^*$. Then, by definition, $\langle \Gamma°, \Theta° \rangle \in L$. But L is closed under substitutivity, so $\langle s(\Gamma°), s(\Theta°) \rangle \in L$. By what we just said, then, $\langle (s\Gamma)°, (s\Theta)° \rangle \in L$. Hence, by the definition of $L^\#$, $\langle s\Gamma, s\Theta \rangle \in L^\#$, as we wanted to show.

Our proof that L* is conservative over L is now easy. Suppose that we know $\langle \Gamma, \Theta \rangle \in L^* \cap \mathbf{P}(\text{Form}) \times \mathbf{P}(\text{Form})$. We then conclude that

PRE-CLASSICAL LOGICS 91

(3) $\quad\langle \Gamma, \Theta\rangle \in L^*;$

(4) $\quad \Gamma, \Theta \subseteq \text{Form}.$

Since $L^* \subseteq L^\#$, (3) implies that $\langle \Gamma, \Theta \rangle \in L^\#$. Hence, by the definition of $L^\#$, $\langle \Gamma^\circ, \Theta^\circ \rangle \in L$. But it follows from (4) that $\Gamma = \Gamma^\circ$ and $\Theta = \Theta^\circ$. Hence, $\langle \Gamma, \Theta \rangle \in L$, as we wanted to show. This proves part (ii) of the theorem.

Part (iii). Obvious.

Part (iv). That L and L* are syntactically equivalent now follows at once from Theorem 2.5.1. ■

COROLLARY 4.2.2. *Two Boolean logics are syntactically equivalent if they have a Boolean extension in common.*

Proof. What is claimed is this: If L_1 and L_2 are Boolean logics in languages $\mathcal{L}_1$ and $\mathcal{L}_2$, respectively, then L_1 and L_2 are syntactically equivalent if there is a logic L* such that the following conditions hold:

(1) $\quad \mathcal{L}^*$ extends both $\mathcal{L}_1$ and $\mathcal{L}_2$;

(2) $\quad$ L* extends both L_1 and L_2.

But this follows at once from the preceding theorem and Corollary 2.5.2. ■

Going back to the definition of immediate Boolean extension above, we may note that the preceding theorem — especially part (iii) — guarantees that we are in agreement with the terminology laid down in Chapter 2 if we refer to L_1 as *the definitional extension of L_0 by* the definition (vi).

If a truth-value function is expressible in a logic, there are in general indefinitely many formulas that could be used as defining formulas in a definition. What difference does it make which one we choose? The answer — or one answer — is given by the next result.

THEOREM 4.2.3. *Let L be a Boolean logic in some language $\mathcal{L}$, and suppose that ϕ is an n-ary truth-value function. Let A_1 and A_2 be any formulas expressing ϕ with respect to some propositional letters $P_0, \ldots, P_{n-1}$. Let ☆ be an n-ary Boolean operator that does not occur in $\mathcal{L}$. Finally, let L_1 and L_2 be the definitional extensions of L by (1) and (2), respectively:*

(1) $\quad ☆[P_0, \ldots, P_{n-1}] \dashv \vdash A_1;$

(2) $\quad ☆[P_0, \ldots, P_{n-1}] \dashv \vdash A_2.$

Then $L_1 = L_2$.

Proof. Since A_1 and A_2 express ϕ with respect to the same propositional letters, it follows by Lemma 4.1.1 that $A_1 \dashv \vdash A_2$ in L. Hence, by (2), condition (1) holds in L_2. But, by definition, L_1 is the smallest logic in which (1) holds. Therefore, $L_1 \subseteq L_2$. By a similar argument, $L_1 \supseteq L_2$. Consequently, $L_1 = L_2$. ∎

4.3. A conservation theorem in pre-classical logic

We say that a Boolean logic L is *truth-value functionally complete* if every truth-value function can be expressed in L. If this is the case, we also say that L is *pre-classical*.

This whole book is a search for a workable exact explication of the intuitive notion of classical propositional logic. Pre-classical logics have so many 'classical' characteristics that one may well ask whether the search should now end. The following theorem is perhaps an argument in favour of doing so.

THEOREM 4.3.1. *Let L be any consistent, pre-classical logic. Let Γ and Θ be sets of purely Boolean formulas. Then $\Gamma \vdash \Theta$ iff $\Gamma \vDash \Theta$.*

Proof. By respect, $\Gamma \vdash \Theta$ if $\Gamma \vDash \Theta$, so we need only prove that also the 'only-if'-part holds.

To begin with we shall assume that the truth-value functions T and F are directly expressed in L — say by the (zeroary) Boolean operators ⊤ and ⊥.

Suppose that $\Gamma \nvdash \Theta$. Then there is some truth-value assignment for the set of Boolean atoms in $\Gamma \cup \Theta$ such that $\bar{f}A = 1$, for all $A \in \Gamma$, while $\bar{f}B = 0$, for all $B \in \Theta$. Since by hypothesis all operators in Γ and Θ are Boolean, the set of Boolean atoms in $\Gamma \cup \Theta$ coincides with the set of propositional letters in $\Gamma \cup \Theta$. For this reason the following condition on a substitution function s is meaningful: If P is a propositional letter occurring in $\Gamma \cup \Theta$, then

$$sP = \top, \quad \text{if } fP = 1,$$
$$= \bot, \quad \text{if } fP = 0.$$

So let s be such a substitution function. (What s does to propositional letters outside $\Gamma \cup \Theta$ is of no concern.) Then, as an inductive argument readily shows, $\bar{f}A = \bar{f}(sA)$, for all Boolean formulas A with propositional letters among those in $\Gamma \cup \Theta$. But, as another inductive argument shows, this means that, for all truth-value assignments g, $\bar{f}A = \bar{g}(sA)$. This implies, for one thing, that

(1) $\vDash A$, for all $A \in s\Gamma$.

For another, it implies that

(2) $B \vDash$, for all $B \in s\Theta$.

Hence, by respect,

(3) $\vdash A$, for all $A \in s\Gamma$,

(4) $\quad\quad\quad\quad\quad\quad\quad\quad$ B $\vdash$, for all B $\in$ sΘ.

In order to prove the theorem we must show that $\Gamma \nvdash \Theta$. Assume the contrary. Then, by substitution,

(5) $\quad\quad\quad\quad\quad\quad\quad\quad$ s$\Gamma \vdash$ sΘ.

By cut, (3) and (5) yield

(6) $\quad\quad\quad\quad\quad\quad\quad\quad$ $\vdash$ sΘ.

By another cut, (4) and (6) yield $\vdash$. That is to say, L is inconsistent, which is a contradiction.

The first part of the proof is now finished: the case when T and F are directly expressed in L. What if they are not? In that case let L* be a Boolean extension of L in which T and F are directly expressible. Let **M** and **M*** be the implicit matrices of L and L*, respectively. Then **M** and **M*** agree on their common domain of definition. For assume not. Let **M**$_*$ be the restriction of **M*** to the set of Boolean operators in the language of L. Let Σ and Ω be any sets of formulas in the language of L. Assume that $\Sigma \vDash^{\mathbf{M}_*} \Omega$. Then also $\Sigma \vDash^{\mathbf{M}^*} \Omega$. Hence, by respect, $\Sigma \vdash_{L^*} \Omega$. By Theorem 4.2.1, L* is conservative over L, so $\Sigma \vdash_L \Omega$. Hence **M**$_*$ respects L. But L is Boolean, so the matrix respecting L is unique (Theorem 3.4.1). Hence **M** = **M**$_*$.

Assume now that $\Gamma \vdash_L \Theta$. Since L* extends L, $\Gamma \vdash_{L^*} \Theta$. Therefore, by what we have just proved in the first part of this proof, $\Gamma \vDash^{\mathbf{M}^*} \Theta$. But Γ and Θ are sets of formulas of the language of L. As we saw, **M** and **M*** agree where they are both defined, so $\Gamma \vDash^{\mathbf{M}} \Theta$. ∎

We may now try to draw the moral of the formal development in the last two sections. Very often we are faced with a pre-classical logic in which some n-ary truth-value function we are interested in is not directly expressible. We see now that there are two ways in which we can proceed in such a case. One is to select a new symbol ☆ for the truth-value function in question and move to a definitional extension in which that truth-value function is directly expressible. The other is to remain in the given logic but pick an appropriate formula A — a Boolean combination of some distinct propositional letters $P_0, \ldots, P_{n-1}$ — and *treat it* as the defining formula in a definition of ☆ of the first kind. In either case the notation '☆$[B_0, \ldots, B_{n-1}]$' would be meaningful, but in rather different ways. In the former case, ☆$[B_0, \ldots, B_{n-1}]$ would be the formula obtained by letting the primitive Boolean operator ☆ operate on the formulas $B_0, \ldots, B_{n-1}$. In the latter, ☆$[B_0, \ldots, B_{n-1}]$ would be a formula obtained by simultaneously substituting B_0 for $P_0, \ldots, B_{n-1}$ for P_{n-1} in the formula A.

The celebrated distinction between use and mention (itself used before but mentioned only now) is helpful in bringing out the difference. We have said it before, but it bears repeating: nowhere in this book do we actually *exhibit* a propositional language — it is always only *described*. In our discussion, the star-shaped typographical symbol '☆' has had two different uses. In both cases, ☆[$B_0, \ldots, B_{n-1}$] is a formula, and '☆[$B_0, \ldots, B_{n-1}$]' is a notation referring to it. Moreover, in both cases it is a matter of convention that '☆[$B_0, \ldots, B_{n-1}$]' refers to ☆[$B_0, \ldots, B_{n-1}$], and that '☆' is part of that notation. The difference is this. In the former ('object-linguistic') case, '☆' is used to refer to a certain primitive symbol (viz., ☆). Not so in the latter ('metalinguistic') case: here '☆' does not refer at all. This is why it is often said, in the latter case, that '☆' is 'an abbreviatory device'.

It is important to realize that '☆[$B_0, \ldots, B_{n-1}$]' does not refer to the same formula in the two cases. In the object linguistic case, ☆ is actually a part of the formula ☆[$B_0, \ldots, B_{n-1}$], which thus is of type ☆ ★ C, for some expression C. In the metalinguistic case, except in one very special, degenerate case, '☆[$B_0, \ldots, B_{n-1}$]' refers to a formula of type △ ★ D, where △ is a primitive Boolean operator, and D is some expression; but in this case, '☆' does not refer to △.

We can now sum up the importance of the formal results. Theorem 4.2.1 assures us that the two ways are equivalent in the sense that they lead, not to identical, but to syntactically equivalent logics. Theorem 4.3.1, in connection with Corollary 3.4.4, establishes that ☆[$P_0, \ldots, P_{n-1}$] expresses the same truth-value function (with respect to $P_0, \ldots, P_{n-1}$) in the two cases. Theorem 4.2.3, finally, shows that even though there is an abundantly wide choice of possible definitions available when new Boolean operators are to be defined, this abundance need not worry us: any definition leads to the same logic.

4.4. The Big Seven again

We now adopt the following convention for Boolean logics L. Whenever one of the following truth-value functions is expressible in L, either directly or indirectly, then we will use the following respective typographical symbol in connection with it:

⊤ in connection with T
⊥ in connection with F
¬ in connection with N
∧ in connection with K,
∨ in connection with A,
→ in connection with C,
↔ in connection with E.

If one of these Big Seven truth-value functions, ϕ say, is expressible in L, and if △ is the symbol to use in connection with ϕ, then we shall say that △ is *available in* L. Thus △ is available in L if and only if ϕ is directly expressible in some Boolean extension of L. Furthermore, for every truth-value assignment f, the following conditions will hold whenever the symbol in question is available (otherwise it is meaningless):

$$\bar{f}(\top) = T.$$
$$\bar{f}(\bot) = F.$$
$$\bar{f}(\neg A) = N(\bar{f}A).$$
$$\bar{f}(A \wedge B) = K(\bar{f}A, \bar{f}B).$$
$$\bar{f}(A \vee B) = A(\bar{f}A, \bar{f}B).$$
$$\bar{f}(A \to B) = C(\bar{f}A, \bar{f}B).$$
$$\bar{f}(A \leftrightarrow B) = E(\bar{f}A, \bar{f}B).$$

Note the following immediate consequence of Theorem 4.1.5.

THEOREM 4.4.1. *If* L *is a Boolean logic, then the following conditions are equivalent:*

(i)   L *is pre-classical*;
(ii) ⊥ *and* → *are available*;
(iii) ¬ *and* → *are available*;
(iv) ¬ *and* ∧ *are available*;
(v)   ¬ *and* ∨ *are available*.

If the preceding discussion seems obscure, an example should be illuminating. Suppose L is a Boolean logic in which → is available. If C is directly expressible in L, then we assume that → is a binary Boolean operator and that **M**(→) = C, where **M** is the implicit matrix. (Our definition of Boolean logic does not rule out the possibility that several distinct operators directly express the same truth-value function.) On the other hand, if C is not directly expressible, then → will be used 'as an abbreviatory device'. Thus suppose that, unlike →, all three of ¬, ∧, and ∨ are primitive in L. Then (cf. section 1.3 for our conventions on how to write formulas) A → B might be the same formula as ¬A ∨ B in which case one can conclude that, for some distinct propositional letters P, Q, we have adopted

$$\neg P \vee Q$$

as our defining formula. Or A → B might be the formula ¬(A ∧ ¬B), in which case we would have adopted, for some distinct P, Q,

$$\neg(P \wedge \neg Q)$$

as our defining formula. Or we might have adopted as our defining formula one of the infinitely many other possible formulas, in which case A → B would be another formula yet.

It is not without interest to rewrite Tables 1 and 2 on pp. 68-9 in the light of our newly adopted conventions. The result is exhibited as Tables 3 and 4. Notice that while the conditions of Tables 1 and 2 are named after the Big Seven truth-value functions, the conditions of Tables 3 and 4 are named after the corresponding typographical symbols.

Table 3
Elimination conditions

EA⊤	*If* *then*	⊤, Γ ⊢ Θ, Γ ⊢ Θ.	EC⊥	*If* *then*	Γ ⊢ Θ, ⊥, Γ ⊢ Θ.
EA¬	*If* *then*	¬A, Γ ⊢ Θ, Γ ⊢ Θ, A.	EC¬	*If* *then*	Γ ⊢ Θ, ¬A, A, Γ ⊢ Θ.
EA∧	*If* *then*	A ∧ B, Γ ⊢ Θ, A, B, Γ ⊢ Θ.	EC∧	*If* *then*	Γ ⊢ Θ, A ∧ B, Γ ⊢ Θ, A *and* Γ ⊢ Θ, B.
EA∨	*If* *then*	A ∨ B, Γ ⊢ Θ, A, Γ ⊢ Θ *and* B, Γ ⊢ Θ.	EC∨	*If* *then*	Γ ⊢ Θ, A ∨ B, Γ ⊢ Θ, A, B.
EA→	*If* *then*	A → B, Γ ⊢ Θ, Γ ⊢ Θ, A *and* B, Γ ⊢ Θ.	EC→	*If* *then*	Γ ⊢ Θ, A → B, A, Γ ⊢ Θ, B.
EA↔	*If* *then*	A ↔ B, Γ ⊢ Θ, A, B, Γ ⊢ Θ *and* Γ ⊢ Θ, A, B.	EC↔	*If* *then*	Γ ⊢ Θ, A ↔ B, A, Γ ⊢ Θ, B *and* B, Γ ⊢ Θ, A.

The conditions of Tables 3 and 4 are not meaningful for logics in general, but, by Theorem 4.4.1, for pre-classical logics they always are. In a sense they even characterize pre-classical logic.

THEOREM 4.4.2. *Let L be a consistent Boolean logic. Then L is pre-classical if and only if L has some Boolean extension in which there are Boolean operators satisfying Tables 3 and 4.*

Proof. The theorem is a corollary of Theorems 3.6.2 and 3.6.3 and the work done in the last two sections. ∎

Tables 3 and 4 have a very familiar look about them: the conditions are simply a well-known kind of Gentzen rules. Among the many standard references in this field we mention Kanger (1957), Kleene (1952), and Takeuti (1975); see also Prawitz (1965).

The system of conditions given in Tables 3 and 4 is by no means independent, by which is meant that some conditions can be derived, in common logics, from the rest. In other words, Theorem 4.4.2 is not the strongest possible result of its kind. More economic systems can easily be given. For example, in a Boolean logic, Table 3 can be derived from Table 4, and conversely. A particularly elegant system has been given by Dana Scott, for example, in his paper (1971).

Table 4
Introduction conditions

IA⊥		⊥ ⊢.	IC⊤		⊢ ⊤.
IA¬	If then	$\Gamma \vdash \Theta, A,$ $\neg A, \Gamma \vdash \Theta.$	IC¬	If then	$A, \Gamma \vdash \Theta,$ $\Gamma \vdash \Theta, \neg A.$
IA∧	If then	$A, B, \Gamma \vdash \Theta,$ $A \wedge B, \Gamma \vdash \Theta.$	IC∧	If then	$\Gamma \vdash \Theta, A$ and $\Gamma \vdash \Theta, B,$ $\Gamma \vdash \Theta, A \wedge B.$
IA∨	If then	$A, \Gamma \vdash \Theta$ and $B, \Gamma \vdash \Theta,$ $A \vee B, \Gamma \vdash \Theta.$	IC∨	If then	$\Gamma \vdash \Theta, A, B,$ $\Gamma \vdash \Theta, A \vee B.$
IA→	If then	$\Gamma \vdash \Theta, A$ and $B, \Gamma \vdash \Theta,$ $A \to B, \Gamma \vdash \Theta.$	IC→	If then	$A, \Gamma \vdash \Theta, B,$ $\Gamma \vdash \Theta, A \to B.$
IA↔	If then	$A, B, \Gamma \vdash \Theta$ and $\Gamma \vdash \Theta, A, B,$ $A \leftrightarrow B, \Gamma \vdash \Theta.$	IC↔	If then	$A, \Gamma \vdash \Theta, B$ and $B, \Gamma \vdash \Theta, A,$ $\Gamma \vdash \Theta, A \leftrightarrow B.$

In effect his system consists of the conditions listed in Table 5 (where Γ and Θ are required to be finite sets). In our terminology Scott's conditions comprise (EA⊤), (EC⊥); (EA¬), (IA¬); (EA∧), (IA∧); (EC∨), (IC∨); (EC→), (IC→); while monotonicity gives the introduction conditions for ⊤ and ⊥.

Seeing Scott's symmetric system, the reader may well wonder at the residue of asymmetry left in it. For example, ∧ is both eliminated and introduced in the antecedent, ∨ both in the consequent. Similarly, ¬ is both eliminated and introduced in the antecedent, → both in the consequent. The answer why this is so is provided by Theorem 3.6.1 and a preference for simple conditions. Evidently it would have been equally acceptable with a condition eliminating and

Table 5
Scott's conditions

S⊤	$\Gamma \vdash \Theta$	iff	$\top, \Gamma \vdash \Theta.$
S⊥	$\Gamma \vdash \Theta$	iff	$\Gamma \vdash \Theta, \bot.$
S¬	$\Gamma \vdash \Theta, A$	iff	$\neg A, \Gamma \vdash \Theta.$
S∧	$A, B, \Gamma \vdash \Theta$	iff	$A \wedge B, \Gamma \vdash \Theta.$
S∨	$\Gamma \vdash \Theta, A, B$	iff	$\Gamma \vdash \Theta, A \vee B.$
S→	$A, \Gamma \vdash \Theta, B$	iff	$\Gamma \vdash \Theta, A \to B.$

introducing $\neg$ in the consequent instead. But in the case of $\wedge$ and $\vee$ the most natural alternative conditions would have been:

(S∧') $\quad\quad\Gamma \vdash \Theta, A$ *and* $\Gamma \vdash \Theta, B\quad$ *iff* $\quad\Gamma \vdash \Theta, A \wedge B$,

(S∨') $\quad\quad A, \Gamma \vdash \Theta$ *and* $B, \Gamma \vdash \Theta\quad$ *iff* $\quad A \vee B, \Gamma \vdash \Theta$.

Undeniably, both are more complicated than Scott's conditions. Similar remarks hold for $\to$. Scott does not give any conditions for $\leftrightarrow$, perhaps because any such condition would have to be more complicated than those in Table 5, but in the style of the alternatives just mentioned we could have:

(S↔) $\quad\quad A, \Gamma \vdash \Theta, B$ *and* $B, \Gamma \vdash \Theta, A\quad$ *iff* $\quad \Gamma \vdash \Theta, A \leftrightarrow B$,

or on the other side:

(S↔') $\quad\quad A, B, \Gamma \vdash \Theta$ *and* $\Gamma \vdash \Theta, A, B\quad$ *iff* $\quad A \leftrightarrow B, \Gamma \vdash \Theta$.

4.5. The Deduction Theorem and the Finiteness Theorem in pre-classical logic

In this section we examine how some well-known theorems appear in the present setting. The Deduction Theorem is a triviality.

THEOREM 4.5.1. (*'Deduction Theorem for Pre-classical Logics.'*) *If* L *is a pre-classical logic, then*

$$A, \Gamma \vdash \Theta, B \quad \textit{iff} \quad \Gamma \vdash \Theta, A \to B.$$

Proof. This is immediate from (EC→) and (IC→). ∎

The following is the traditional form of the Deduction Theorem, obtained when $\Theta = \emptyset$.

COROLLARY 4.5.2. *If* L *is a pre-classical logic, then*

$$A, \Gamma \vdash B \quad \textit{iff} \quad \Gamma \vdash A \to B.$$

Before proceeding we shall introduce a few simplifying conventions. Note that $A_0 \wedge (A_1 \wedge A_2) \dashv\vdash (A_0 \wedge A_1) \wedge A_2$ in any Boolean logic in which $\wedge$ is available. In general all ways of inserting parentheses into the expression

(#) $\quad\quad\quad\quad\quad\quad\quad A_0 \wedge ... \wedge A_{m-1}$

yield interdeducible formulas, even if $m > 2$. For this reason we shall usually omit parentheses in such contexts and recognize expressions of type (#) as meaningful, even though strictly speaking they are not; the understanding is that, in a particular case, any way of restoring parentheses will be acceptable. Expressions of type

(♭) $\quad\quad\quad\quad\quad\quad\quad B_0 \vee ... \vee B_{n-1}$

are treated similarly. These conventions are in common use and increase both convenience and generality.

For the limiting cases of (#) and (♭) — the 'empty conjunction' and the 'empty disjunction', respectively — we adopt two further conventions:

If m = 0, then (#) reduces to $\top$.

If n = 0, then (♭) reduces to $\bot$.

Again, those are common conventions.

The following lemma contains generalized versions of (**EA**∧) and (**IA**∧) on the one hand and of (**EC**∨) and (**IC**∨) on the other. The two parts are easily proved by induction (on m and n, respectively), so we omit the proofs.

LEMMA 4.5.3. *In any pre-classical logic,*

(i) $A_0, \ldots, A_{m-1}, \Gamma \vdash \Theta$ *iff* $A_0 \wedge \ldots \wedge A_{m-1}, \Gamma \vdash \Theta$;

(ii) $\Gamma \vdash \Theta, B_0, \ldots, B_{n-1}$ *iff* $\Gamma \vdash \Theta, B_0 \vee \ldots \vee B_{n-1}$.

We are now able to state the following important result:

THEOREM 4.5.4. ('*Finiteness Theorem for Finitary Pre-classical Logics.*') *If* L *is a finitary pre-classical logic, then* $\Gamma \vdash \Theta$ *if and only if there are* m, n $\geq$ 0 *and formulas* $A_0, \ldots, A_{m-1} \in \Gamma$ *and* $B_0, \ldots, B_{n-1} \in \Theta$ *such that*

$$\vdash A_0 \wedge \ldots \wedge A_{m-1} \to B_0 \vee \ldots \vee B_{n-1}.$$

Proof. Let us first remark that the displayed formula is meant to be unambiguous: $\to$ is the main operator.

Suppose now that L is a finitary pre-classical logic and that $\Gamma \vdash \Theta$. Hence, there exist finite subsets $\Gamma^\circ \subseteq \Gamma$, $\Theta^\circ \subseteq \Theta$ such that $\Gamma^\circ \vdash \Theta^\circ$. Say $\Gamma^\circ = \{A_0, \ldots, A_{m-1}\}$ and $\Theta^\circ = \{B_0, \ldots, B_{n-1}\}$. Then

$$A_0, \ldots, A_{m-1} \vdash B_0, \ldots, B_{n-1}.$$

L is pre-classical, so, by Lemma 4.5.3, $A_0 \wedge \ldots \wedge A_{m-1} \vdash B_0 \vee \ldots \vee B_{n-1}$. The Deduction Theorem then yields the desired conclusion.

To prove the converse, assume that, for some $A_0, \ldots, A_{m-1} \in \Gamma$ and $B_0, \ldots, B_{n-1} \in \Theta$, we have $\vdash A_0 \wedge \ldots \wedge A_{m-1} \to B_0 \vee \ldots \vee B_{n-1}$. By the Deduction Theorem, $A_0 \wedge \ldots \wedge A_{m-1} \vdash B_0 \vee \ldots \vee B_{n-1}$. Hence, again by Lemma 4.5.3,

$$A_0, \ldots, A_{n-1} \vdash B_0, \ldots, B_{n-1}.$$

Therefore, by monotonicity, $\Gamma \vdash \Theta$. ∎

Note that the condition that L is pre-classical is used in both halves of the proof, whereas the condition that L is finitary is used only once. A more traditional form of the preceding theorem is the following.

COROLLARY 4.5.5. *If* L *is a finitary pre-classical logic, then* $\Gamma \vdash B$ *if and only if, for some* $m \geq 0$, *there are* $A_0, \ldots, A_{m-1} \in \Gamma$ *such that*

$$\vdash A_0 \wedge \ldots \wedge A_{m-1} \to B.$$

THEOREM 4.5.6. *Let* L *and* L' *be any finitary pre-classical logics in the same language. Then* $\text{Th}(L) = \text{Th}(L')$ *if and only if* $L = L'$.

Proof. Assume that $\text{Th}(L) = \text{Th}(L')$. Take any Γ, Θ such that $\Gamma \vdash_L \Theta$. Then, by the Finiteness Theorem, there are some $A_0, \ldots, A_{m-1} \in \Gamma$ and $B_0, \ldots, B_{n-1} \in \Theta$ such that the formula

$$A_0 \wedge \ldots \wedge A_{m-1} \to B_0 \vee \ldots \vee B_{n-1}$$

is a thesis of L. By assumption, then, it is a thesis of L' as well. Hence, by another application of the Finiteness Theorem, $\Gamma \vdash_{L'} \Theta$. This argument shows that $L \subseteq L'$. A similar argument shows that $L \supseteq L'$, and hence $L = L'$. The converse is obvious. ■

The significance of the last theorem resides in the fact that it shows that finitary pre-classical logics are — in a certain sense — reducible to their sets of theses. When we study such logics, we lose nothing if we identify them with their respective sets of theses; which is not to say that it is always a good idea to do so.

4.6. Makinson's Warning

Among the many possible conditions on logics that there are we shall now focus on one that is rarely mentioned but often taken for granted. It is the condition that the logic be insensitive to the choice of primitive Boolean operators: that it somehow does not matter what primitive Boolean operators there are. For instance, in classical logic authors sometimes do not bother to describe their object language in detail, requiring only that it contain some truth-functionally complete set of Boolean operators. Thereby they implicitly assert that whatever results they achieve are valid for every language that meets the requirement. But they never *prove* that this is so.

Are pre-classical logics sensitive to the choice of primitive Boolean operators? Theorems 4.2.1 and 4.2.3 would lead one to believe that the answer is negative: in such a logic every truth-value function is expressible, whether or not directly so would not seem to matter. Nevertheless, pre-classical logics do suffer from a kind of sensitivity, and in a slightly surprising manner. This was evidently first observed by David Makinson in 1973. We shall devote this section to working out an example originally due to him. We first prove an auxiliary result.

LEMMA 4.6.1. *Let L be the smallest pre-classical logic in some language $\mathcal{L}$. Let Δ be any set of formulas of $\mathcal{L}$ such that, for all substitution functions s, $\Delta \vDash sA$, for all $A \in \Delta$. Define*

$$L\Delta = \{\langle \Gamma, \Theta \rangle : \Delta, \Gamma \vDash \Theta\},$$

where Γ and Θ range over sets of formulas of $\mathcal{L}$. Then $L\Delta$ is the smallest pre-classical logic such that $\vdash A$ holds for all $A \in \Delta$.

Proof. The problematic part of the proof that $L\Delta$ is pre-classical is the subproof that $L\Delta$ is closed under substitution. All the other parts follow from Theorem 3.2.1 and by the observation that assertivity and truth-value functional completeness are inherited from L.

Take any $\langle \Gamma, \Theta \rangle \in L\Delta$, and any substitution function s. Then $\Delta, \Gamma \vDash \Theta$, and so, by Theorem 3.2.1, $s\Delta, s\Gamma \vDash s\Theta$. By hypothesis, $\Delta \vDash A$, for all $A \in s\Delta$. Consequently, by Theorem 3.2.1 (the cut part), $\Delta, s\Gamma \vDash s\Theta$, and so $\langle s\Gamma, s\Theta \rangle \in L\Delta$. This proves that $L\Delta$ satisfies the condition (Subst), and so we conclude that $L\Delta$ is indeed pre-classical. (So $L\Delta$ is the smallest pre-classical logic to extend both L and $\{\langle \emptyset, A \rangle : A \in \Delta\}$; this lends some justification to the name '$L\Delta$'.)

It will now be enough to prove that if L' is any pre-classical logic (in $\mathcal{L}$) such that, for all $A \in \Delta$, $\vdash A$, then $L\Delta \subseteq L'$. Suppose that $\langle \Gamma, \Theta \rangle \in L\Delta$. Then $\Delta, \Gamma \vDash \Theta$. Hence, by respect, $\Delta, \Gamma \vdash_L \Theta$. Since L is the smallest pre-classical logic, $L \subseteq L'$, and so $\Delta, \Gamma \vdash_{L'} \Theta$. But L' is closed under cut, so, using the fact that, for all $A \in \Delta$, $\vdash A$, we conclude that $\Gamma \vdash_{L'} \Theta$. ∎

COROLLARY 4.6.2. *Let L be the smallest pre-classical logic (in a given language). Then*

$$L = \{\langle \Gamma, \Theta \rangle : \Gamma \vDash \Theta\}.$$

Proof. For every substitution function s, $s\emptyset = \emptyset$. Hence the desired result follows trivially from the preceding lemma by putting $\Delta = \emptyset$. ∎

Let $\mathcal{L}_1 = \langle \text{Lett}, \text{Bop}_1, \text{Iop}, r_1 \rangle$ and $\mathcal{L}_2 = \langle \text{Lett}, \text{Bop}_2, \text{Iop}, r_2 \rangle$ be two languages such that

$$\text{Bop}_1 = \{\bot, \rightarrow\},$$
$$\text{Bop}_2 = \{\neg, \rightarrow\},$$
$$\text{Iop} = \{\Box\},$$
$$r_1(\bot) = 0,$$

$$r_2(\neg) = 1,$$
$$r_1(\to) = r_2(\to) = 2,$$
$$r_1(\Box) = r_2(\Box) = 1.$$

(Actually, some of the information about r_1 and r_2 is redundant, for $\bot$, $\neg$, and $\to$ are as specified in section 4.5.) Let L_1 and L_2 be the smallest pre-classical logics in $\mathcal{L}_1$ and $\mathcal{L}_2$, respectively. It is not very surprising that L_1 and L_2 turn out to be syntactically equivalent. But it may be instructive to give the proof.

THEOREM 4.6.3. *L_1 and L_2 are syntactically equivalent.*

Proof. By Corollary 4.2.2 it will be enough to show that L_1 and L_2 have a Boolean extension in common.

Let P be a fixed propositional letter. (Such a P is always available, for by general assumption Lett is denumerably infinite and hence certainly non-empty.) Let L_1^* be the definitional extension of L_1 by the obvious definition

(1) $\qquad\qquad\qquad \neg P \dashv\vdash P \to \bot.$

Similarly, let L_2^* be the definitional extension of L_2 by the equally obvious definition

(2) $\qquad\qquad\qquad \bot \dashv\vdash \neg(P \to P).$

It is clear that L_1^* and L_2^* are logics in the same language, viz., $\langle$Lett, $\text{Bop}_1 \cup \text{Bop}_2$, Iop, $r_1 \cup r_2\rangle$. By Theorem 4.2.1, L_1^* is conservative over L_1, and L_2^* is conservative over L_2. Hence in order to prove the theorem it is enough to show that $L_1^* = L_2^*$.

By definition, L_1^* is the smallest common logic extending L_1 to contain (1). But L_2^* is certainly common (in fact, pre-classical), and (1) holds also in L_2^*. Hence $L_1^* \subseteq L_2^*$.

By an analogous argument, $L_1^* \supseteq L_2^*$. Hence $L_1^* = L_2^*$. ∎

So L_1 and L_2 are syntactically equivalent. Yet there is one interesting difference between them, as the following theorems show.

THEOREM 4.6.4. *L_1 is the intersection of two of its proper pre-classical extensions in $\mathcal{L}_1$.*

Proof. Using the notational convention introduced in Lemma 4.6.1, we define

$$L_{11} = L_1\{\Box\bot\}, \text{ and } L_{21} = L_2\{\Box\bot \to \bot\}.$$

It is easy to see that both L_{11} and L_{21} are proper extensions of L_1. We claim that

$$L_1 = L_{11} \cap L_{12}.$$

That $L_1 \subseteq L_{11} \cap L_{12}$ is clear. To prove the converse, suppose that $\langle \Gamma, \Theta \rangle \in L_{11} \cap L_{12}$. (With so many subscripts, the longer notation seems more suitable than the turnstyle variety.) Then $\langle \Gamma, \Theta \rangle \in L_{11}$ and $\langle \Gamma, \Theta \rangle \in L_{12}$, and so, by Lemma 4.6.1,

(1) $\qquad\qquad\qquad \Box\bot, \Gamma \vDash^{M_1} \Theta$,

(2) $\qquad\qquad\qquad \Box\bot \to \bot, \Gamma \vDash^{M_1} \Theta$,

where M_1 is the matrix implicit in L_1. Let f be any assignment of truth-values to the Boolean atoms of $\mathcal{L}_1$ such that, for all $A \in \Gamma$, $f^{M_1}(A) = 1$. It is clear that either $f^{M_1}(\Box\bot) = 1$ or $f^{M_1}(\Box\bot \to \bot) = 1$. Hence, by whichever of (1) and (2) that applies, $f^{M_1}(B) = 1$, for some $B \in \Theta$. This proves that $\Gamma \vDash^{M_1} \Theta$. Hence, by respect, $\langle \Gamma, \Theta \rangle \in L_1$, as we wanted to show. ∎

THEOREM 4.6.5. L_2 *is not the intersection of any two of its proper pre-classical extensions in* $\mathcal{L}_2$.

Proof. Assume that L_{21} and L_{22} are pre-classical extensions (in $\mathcal{L}_2$) of L_2, and that both extensions are proper. Then there are sets $\Gamma_1, \Gamma_2, \Theta_1, \Theta_2$ such that

(1) $\qquad\qquad\qquad \langle \Gamma_1, \Theta_1 \rangle \in L_{21}$,

(2) $\qquad\qquad\qquad \langle \Gamma_2, \Theta_2 \rangle \in L_{22}$,

(3) $\qquad\qquad\qquad \langle \Gamma_1, \Theta_1 \rangle \notin L_2$,

(4) $\qquad\qquad\qquad \langle \Gamma_2, \Theta_2 \rangle \notin L_2$.

It is not excluded that $\Gamma_1 \cup \Theta_1$ and $\Gamma_2 \cup \Theta_2$ have some Boolean atom in common. Therefore we proceed as follows. As remarked above, it is a general assumption that our language contains denumerably many propositional letters. Let $P_0, P_1, \ldots, P_n, \ldots$ be an exhaustive enumeration of them in which every propositional letter occurs only once. We define two substitution functions as follows: for all n,

$$s_1 P_n = P_{2n+1}, \text{ and } s_2 P_n = P_{2n}.$$

Define $\Gamma_i' = s_i \Gamma$ and $\Theta_i' = s_i \Theta$ (i = 1, 2). It is clear that $\Gamma_1' \cup \Theta_1'$ and $\Gamma_2' \cup \Theta_2'$ have no propositional letter in common. Therefore, since $\mathcal{L}_2$ does not contain any propositional constant, $\Gamma_1' \cup \Theta_1'$ and $\Gamma_2' \cup \Theta_2'$ also have no Boolean atom in common. Moreover, since pre-classical logics are closed under substitution, (1) and (2) imply

$$\langle \Gamma_1', \Theta_1' \rangle \in L_{21}, \text{ and } \langle \Gamma_2', \Theta_2' \rangle \in L_{22}.$$

Hence, by monotonicity,

$$\langle \Gamma_1' \cup \Gamma_2', \Theta_1' \cup \Theta_2' \rangle \in L_{21}, \text{ and } \langle \Gamma_1' \cup \Gamma_2', \Theta_1' \cup \Theta_2' \rangle \in L_{22},$$

and so

(5) $\langle \Gamma'_1 \cup \Gamma'_2, \Theta'_1 \cup \Theta'_2 \rangle \in L_{21} \cap L_{22}$.

By Corollary 4.6.2, for all sets Ξ and Π.

(6) $\langle \Xi, \Pi \rangle \in L_2$ iff $\Xi \vDash^{M_2} \Pi$,

where M_2 is the matrix implicit in L_2. Hence, by (3) and (4), $\Gamma_i \nvDash^{M_2} \Theta_i$, for $i = 1, 2$; and it is easy to see that this implies that

$\Gamma'_1 \nvDash^{M_2} \Theta'_1$, and $\Gamma'_2 \nvDash^{M_2} \Theta'_2$.

But then there are truth-value assignments f_1 and f_2 for the set of Boolean atoms such that

$$f_1^{M_2}(A) = 1, \text{ for all } A \in \Gamma'_1,$$
$$f_2^{M_2}(A) = 1, \text{ for all } A \in \Gamma'_2,$$
$$f_1^{M_2}(B) = 0, \text{ for all } B \in \Theta'_1,$$
$$f_2^{M_2}(B) = 0, \text{ for all } B \in \Theta'_2.$$

Since $\Gamma'_1 \cup \Theta'_1$ and $\Gamma'_2 \cup \Theta'_2$ have no Boolean atom in common, the following defines a truth-value assignment g for a certain subset of Boolean atoms: if C is a Boolean atom, then

$gC = f_1(C)$, if C contains only odd-numbered propositional letters,

$= f_2(C)$, if C contains only even-numbered propositional letters.

It is readily proved, then, that $g^{M_2}(A) = 1$, for all $A \in \Gamma'_1 \cup \Gamma'_2$, and $g^{M_2}(B) = 0$, for all $B \in \Theta'_1 \cup \Theta'_2$. Consequently, $\Gamma'_1, \Gamma'_2 \nvDash^{M_2} \Theta'_1, \Theta'_2$. By (6), therefore,

(7) $\langle \Gamma'_1 \cup \Gamma'_2, \Theta'_1 \cup \Theta'_2 \rangle \notin L_2$.

From (5) and (7) the theorem follows. ∎

Here we are: L_1 and L_2 are syntactically equivalent, yet they differ in the manner described by Theorems 4.6.4 and 4.6.5. Using two popular philosophical terms, we may say that even though L_1 and L_2 have the same 'internal' properties, they do not share all the 'external' ones. At least at first sight this is a disturbing state of affairs which calls for some reflection. We shall refer to this discovery as 'Makinson's Warning', since the word 'warning' was used by Makinson himself.

There are several ways to react to Makinson's Warning. One is to go on finding it disturbing and accept, as a fact of life, that logic is language-sensitive in the way we have just seen. Another would be to eliminate the problem by legislation; say, by requiring all languages to contain ⊥ as a primitive Boolean operator, or, at the other extreme, by banning languages that contain propositional constants. Yet another would be to avoid the problem by confining attention to a restricted family of pre-classical logics to which Makinson's Warning does not apply. Perhaps there are still others. The question, which way out to choose, is one of definition — we cannot say that it would be wrong to take one of the first two. However, here we will go in what we regard as the most interesting and natural way: the third way. The restricted family of pre-classical logics that we have in mind is that of *classical logics*, a notion to which we shall devote the following chapter.

5
CLASSICAL LOGICS

5.1. Replacement of tautological equivalents: formulas

Before the notion of classical logic can be defined, several technical concepts must be introduced. Let $\mathcal{L}$ be a propositional language and $\mathbf{M}$ a matrix for $\mathcal{L}$. Let A and B be formulas in $\mathcal{L}$. Then we say that B *can be obtained from* A *in at most one step by replacement of* **M**-*equivalents*, in symbols $A \underset{\mathbf{M}}{\sim} B$, if there are formulas C, $\overline{C}$ and expressions E, F, not necessarily non-empty, such that the following conditions are satisfied:

(i) $\qquad A = E \star C \star F,$

(ii) $\qquad B = E \star \overline{C} \star F,$

(iii) $\qquad C \dashv \vdash^{\mathbf{M}} \overline{C}.$

See section 1.2 for the $\star$-notation. Do not confuse replacement of tautological equivalents with replacement of interdeducible formulas — see section 2.3.

The notation $A \underset{\mathbf{M}}{\sim}^n B$ has the meaning that is customary with binary relations. Thus, in particular, $A \underset{\mathbf{M}}{\sim}^0 B$ if and only if $A = B$. If $A \underset{\mathbf{M}}{\sim}^n B$ we say that B *can be obtained from* A *in at most* n *steps by replacement of* **M**-*equivalents*. It is at once clear that $\underset{\mathbf{M}}{\sim}$ is a relation over the set of formulas in $\mathcal{L}$ that is reflexive, symmetric but not transitive. The ancestral $\underset{\mathbf{M}}{\sim}^*$ of $\underset{\mathbf{M}}{\sim}$ — that is, the reflexive, transitive closure of $\underset{\mathbf{M}}{\sim}$ — is of course an equivalence relation. If $A \underset{\mathbf{M}}{\sim}^* B$, then we say that B can be obtained from A *by replacement of* **M**-*equivalents*.

If L is a Boolean logic in $\mathcal{L}$ and $\mathbf{M}$ its implicit matrix, we may speak of 'tautological equivalents' instead of '$\mathbf{M}$-equivalents'. In this case, the reference to $\mathbf{M}$ may be omitted in the notation: thus $A \sim B$ if $A \underset{\mathbf{M}}{\sim} B$, etc. The new relation $\underset{\mathbf{M}}{\sim}$ is of course related to the old relation $\dashv\vdash^{\mathbf{M}}$:

THEOREM 5.1.1. *Let* A *and* B *be purely Boolean formulas. Then* $A \underset{\mathbf{M}}{\sim} B$ *iff* $A \dashv\vdash^{\mathbf{M}} B$.

Proof. **If-part.** Suppose that $A \dashv\vdash^{\mathbf{M}} B$. Then just note that $A = \emptyset \star A \star \emptyset$, and

$B = \emptyset \star B \star \emptyset$. So for this part of the theorem, the assumption that A and B are purely Boolean is not needed.

Only-if-part. Suppose that $A \underset{M}{\sim} B$. Then there are formulas $C, \overline{C}$ and expressions X, Y such that $A = X \star C \star Y$, and $B = X \star \overline{C} \star Y$, where $C \dashv\vdash^M \overline{C}$. The proof now proceeds by induction on the complexity of A.

If A = P, for some propositional letter P, then $X = Y = \emptyset$, and so A = C, $B = \overline{C}$, whence $A \dashv\vdash^M B$.

Suppose then that $A = \star[D_0, \ldots, D_{n-1}]$, for some n-ary Boolean operator $\star$ and some formulas $D_0, \ldots, D_{n-1}$ such that — and this is the induction hypothesis — the only-if-part of the theorem holds for them.

If $X = \emptyset$, then, by Lemma 1.3.1, $Y = \emptyset$. Thus, just as before, A = C, $B = \overline{C}$, and $A \dashv\vdash^M B$.

If $X \neq \emptyset$, then, using Lemma 1.4.2 and Theorem 1.4.3, we conclude that there is a unique $k < n$ as well as formulas $D'_0, \ldots, D'_{n-1}$ such that the following conditions hold: $B = \star[D'_0, \ldots, D'_{n-1}]$, C is a subformula of D_k, $\overline{C}$ is a subformula of D'_k, and for all $i < n$, if $i \neq k$ then $D_i = D'_i$. By the induction hypothesis, $D_k \dashv\vdash^M D'_k$. Since $\star$ is Boolean, it follows that $A \dashv\vdash^M B$. ∎

We list the next result for future reference. The proof of the only-if-part of Theorem 5.1.1 was given in detail because it is a typical proof by induction when the relation $\underset{M}{\sim}$ is involved. The proof of this lemma is quite analogous.

LEMMA 5.1.2. *Let A, B be any formulas and s any substitution function. Then $A \underset{M}{\sim} B$ implies* $sA \underset{M}{\sim} sB$.

The following is an obvious but important auxiliary result.

LEMMA 5.1.3. *If $\oplus$ is any n-ary operator, and $A_i \underset{M}{\sim}^* B_i$, for all $i < n$, then $\oplus[A_0, \ldots, A_{n-1}] \underset{M}{\sim}^* \oplus[B_0, \ldots, B_{n-1}]$.*

Proof. Informally it is not difficult to see that the assertion is true. For by assumption, for each $i < n$, $A_i \underset{M}{\sim}^{m_i} B_i$, for some m_i. So each B_i can be obtained from A_i in m_i steps by replacement of M-equivalents. Hence $\oplus[B_0, \ldots, B_{n-1}]$ can be obtained from $\oplus[A_0, \ldots, A_{n-1}]$ in

$$\sum_{i=0}^{n-1} m_i$$

steps by replacement of M-equivalents. A more formal proof can be given by induction on the sum of the numbers of steps. ∎

5.2. Replacement of tautological equivalents: sets of formulas

It is desirable to generalize the relations $\underset{M}{\sim}$ and $\underset{M}{\sim}^*$ to sets of formulas. One definition that suggests itself if the following: $\Gamma \underset{M}{\sim} \Delta$ if there are $A \in \Gamma$ and $B \in \Delta$ such that

(i) $\qquad\qquad\qquad A \underset{M}{\sim} B;$

(ii) $\qquad\qquad\qquad \Gamma - \{A, B\} = \Delta - \{A, B\}.$

The new relation $\underset{M}{\sim}$ is reflexive and symmetric, but, like the one between formulas, again not transitive. As before, the ancestral relation $\underset{M}{\sim}^*$ is an equivalence relation. However, in the context of sets of formulas, the relation $\underset{M}{\sim}^*$ so defined is not quite what we want. The reason has to do with cardinality: a set can be related by $\underset{M}{\sim}^*$ only if one has been obtained from the other by replacement of M-equivalents in at most a finite number of steps. But intuitively we would sometimes like to consider two sets equivalent where infinitely many formulas have been changed.

For example, take any pre-classical logic and define $\text{Lett}' = \{P \wedge P : P \in \text{Lett}\}$. Then intuition would have it that Lett and Lett' are equivalent as far as replacement of tautological equivalents goes. But since Lett is infinite we cannot deduce Lett $\sim^*$ Lett'.

It seems that the relation that we do want — let it be denoted by $\underset{M}{\approx}$ — should hold between two sets Γ and Δ if (i) every formula of Γ either is in Δ, or else some other formula is that was obtained from the first one in at most finitely many steps by replacement of M-equivalents, and (ii) every formula of Δ either is in Γ, or else it was obtained from one in at most finitely many steps by replacement of M-equivalents. Therefore, let us adopt as a formal definition that $\Gamma \underset{M}{\approx} \Delta$ if and only if both the following conditions are satisfied:

(i') $\qquad\qquad\qquad \forall A \in \Gamma \; \exists B \in \Delta \; (A \underset{M}{\sim}^* B);$

(ii') $\qquad\qquad\qquad \forall B \in \Delta \; \exists A \in \Gamma \; (A \underset{M}{\sim}^* B).$

LEMMA 5.2.1. $\Gamma \underset{M}{\approx} \Delta$ *if and only if there are functions* $f : \Gamma \to \Delta$ *and* $g : \Delta \to \Gamma$ *such that, for all* $A \in \Gamma$ *and* $B \in \Delta$,

(i'') $\qquad\qquad\qquad A \underset{M}{\sim}^* fA;$

(ii'') $\qquad\qquad\qquad B \underset{M}{\sim}^* gB.$

Proof. By the Axiom of Choice, if necessary. ∎

THEOREM 5.2.2. $\underset{M}{\approx}$ *is an equivalence relation.*

Proof. Follows easily from the lemma. ∎

The replacement relations we have discussed are interesting mainly in connection with the following three conditions, meaningful for all logics that are at least Boolean. (The reader is reminded of our convention to allow 'M' to be dropped in our notation if it denotes the implicit matrix of a Boolean logic under discussion, and it is clear which logic is intended.)

(RTE$_1$)   If $A \sim B$, then $A \dashv\vdash B$.

(RTE$_2$)   If $A \sim^* B$, then $A \dashv\vdash B$.

(RTE$_3$)   If $\Gamma \approx \Gamma'$ and $\Theta \approx \Theta'$, then $\Gamma \vdash \Theta$ implies that $\Gamma' \vdash \Theta'$.

Each of these conditions may be read as a requirement that L be closed under replacement of tautological equivalents. In view of the following theorem there is usually no need for us to distinguish between them. When we refer to them collectively, we use the notation '(RTE)', without subscript.

THEOREM 5.2.3. *For Boolean logics, conditions* (RTE$_1$), (RTE$_2$), *and* (RTE$_3$) *are equivalent.*

Proof. Let L be a Boolean logic. We will establish the chain (RTE$_1$) implies (RTE$_2$) implies (RTE$_3$) implies (RTE$_1$).

(RTE$_1$) implies (RTE$_2$). It is enough to prove the claim that, for all A and B, whenever $A \sim^n B$, then $A \dashv\vdash B$. This claim is proved by induction on n.

The claim is obviously true for $n = 0$. Suppose then that it is true for n (the induction hypothesis), and that $A \sim^{n+1} B$. Then there is some C such that $A \sim^n C$ and $C \sim B$. By the induction hypothesis, $A \dashv\vdash C$, and by (RTE$_1$) $C \dashv\vdash B$. By cut, then, $A \dashv\vdash B$.

(RTE$_2$) implies (RTE$_3$). Assume that $\Gamma \approx \Gamma'$ and $\Theta \approx \Theta'$, and that

(1) $\quad\quad\quad\quad\quad\quad \Gamma \vdash \Theta.$

Take any $A \in \Gamma$. Then, since $\Gamma \approx \Gamma'$, there is some $B \in \Gamma'$ such that $B \sim^* A$. Hence, by (RTE$_2$), $B \vdash A$. By monotonicity, $\Gamma' \vdash A$. Thus we have proved that

(2) $\quad\quad\quad\quad \Gamma' \vdash A$, for every $A \in \Gamma$.

By the same token,

(3) $\quad\quad\quad\quad C \vdash \Theta'$, for every $C \in \Theta$.

By cut we infer from (1) and (2) that $\Gamma' \vdash \Theta$. Similarly, from this and (3) we infer, by cut, that $\Gamma' \vdash \Theta'$.

(RTE$_3$) implies (RTE$_1$). Suppose that $A \sim B$. Trivially, $A \sim A$. Hence

$\{A\} \approx \{A\}$ and $\{A\} \approx \{B\}$. But, by reflexivity, $A \vdash A$. By (RTE$_3$), therefore, $A \vdash B$. ∎

5.3. Classical operators

Let L be any Boolean logic. We say that an n-ary operator $\oplus$ is *classical* (*in* L) if the following condition holds for all $A_0, \ldots, A_{n-1}, B_0, \ldots, B_{n-1}$:

if $A_i \sim B_i$, for all $i < n$, then $\oplus[A_0, \ldots, A_{n-1}] \dashv\vdash \oplus[B_0, \ldots, B_{n-1}]$.

COROLLARY 5.3.1. *Zeroary operators are classical.*

Proof. The claim is vacuously true. ∎

Formally, the concept of classicalness is somewhat similar to that of congruence (section 2.3). It is natural to ask whether they are comparable. The answer, given by the next two theorems, is that they are not.

THEOREM 5.3.2. *Not all classical operators are necessarily congruential in a pre-classical logic.*

Proof. Let $\mathcal{L}$ be a language with $\bot$ and $\to$ among its Boolean operators, and with the unary operator $\triangledown$ as the sole non-Boolean one. We define the following non-standard semantics. Let M be a matrix for $\mathcal{L}$ (whence, by our conventions, $M(\bot) = F$ and $M(\to) = C$). Let f be any truth-value assignment for the set of Boolean atoms. Then we write $\bar{f}$ for f^M in the usual way. But we also introduce a new function $\bar{\bar{f}}$ on the set of all formulas as follows. If P is a propositional letter, then $\bar{\bar{f}}P = fP$. If ☆ is an n-ary Boolean operator, then

$$\bar{\bar{f}}(\text{☆}(A_0, \ldots, A_{n-1})) = M(\text{☆})(\bar{\bar{f}}A_0, \ldots, \bar{\bar{f}}A_{n-1}).$$

Finally, $\bar{\bar{f}}(\triangledown A) = \bar{f}A$. Let us call a truth-value assignment f 'special' if, for all A, $f(\triangledown A) = 1$. Define $\Gamma \Vdash^\dagger \Theta$ to hold if and only if, for all special truth-value assignments f for the set of Boolean atoms, if $\bar{\bar{f}} A = 1$ for all $A \in \Gamma$, then $\bar{\bar{f}} B = 1$ for some $B \in \Theta$. Let $L^\dagger$ be the set of all pairs $\langle \Gamma, \Theta \rangle$ such that $\Gamma \Vdash^\dagger \Theta$.

Let f be any special truth-value assignment for the set of Boolean atoms. Then

$$\bar{\bar{f}}(\bot) = 0,$$
$$\bar{\bar{f}}(\triangledown\bot) = \bar{f}(\bot) = 0,$$
$$\bar{\bar{f}}(\triangledown\triangledown\bot) = \bar{f}(\triangledown\bot) = f(\triangledown\bot) = 1.$$

In $L^\dagger$, therefore, $\bot \dashv\vdash \triangledown\bot$ but not $\triangledown\bot \dashv\vdash \triangledown\triangledown\bot$, so $\triangledown$ is not congruential in $L^\dagger$. It would only remain to show that

CLASSICAL LOGICS   111

(†) $L^\dagger$ is pre-classical, and

(††) ∇ is classical in $L^\dagger$.

Proof of (†). The proof of (†) is similar to that of Theorem 3.2.1. As with the latter, the main difficulty is with checking out substitution. We sketch this part of the proof as an example.

First suppose that s is any substitution function and f and g any truth-value assignments for the set of Boolean atoms such that gC = fsC, for every Boolean atom C. Then, we claim,

(1) $\bar{g}A = \bar{f}sA$ and $\bar{\bar{g}}A = \bar{\bar{f}}sA$, for every formula A;

(2)   If f is special, then so is g.

Claim (1) is proved by simultaneous induction on A. Claim (2) is obvious; if f is special, then $g(\nabla A) = f(s\nabla A) = f(\nabla sA) = 1$.

Suppose now that $s\Gamma \not\Vdash^\dagger s\Theta$, for some substitution function s. Then there is some special truth-value assignment f for the set of Boolean atoms such that

$$\bar{\bar{f}}A = 1, \text{ for all } A \in s\Gamma, \text{ and } \bar{\bar{f}}B = 0, \text{ for all } B \in s\Theta.$$

Hence

$$\bar{\bar{fs}}A = 1, \text{ for all } A \in \Gamma, \text{ and } \bar{\bar{fs}}B = 0, \text{ for all } B \in \Theta.$$

Define gC = fsC, for all Boolean atoms C. Then, by (1),

$$\bar{\bar{g}}A = 1, \text{ for all } A \in \Gamma, \text{ and } \bar{\bar{g}}B = 0, \text{ for all } B \in \Theta.$$

By (2), g is special. Hence $\Gamma \not\Vdash^\dagger \Theta$. So $L^\dagger$ is closed under substitution.

Proof of (††). Also this part of the proof we only sketch. First notice that, thanks to (†), the notation $A \sim B$ is now meaningful. Next observe that if f is a special truth-value assignment for the set of Boolean atoms, then $A \sim B$ implies that $\bar{f}A = \bar{f}B$. A rigorous proof of this assertion can be given along the lines of the proof of Theorem 5.1.1.

The proof of (††) is now easy. Suppose that $A \sim B$. Let f be any special truth-value assignment for the set of Boolean atoms. Then, by what was just asserted, $\bar{f}(\nabla A) = \bar{f}A = \bar{f}B = \bar{f}(\nabla B)$. Hence, in $L^\dagger$, $\nabla A \dashv \vdash \nabla B$, the desired result. ∎

THEOREM 5.3.3. *Not all congruential operators are necessarily classical in a pre-classical logic.*

Proof. Let L be the smallest pre-classical logic in a language that contains a non-Boolean, unary operator △. Assume that ⊥, ¬, ∧ are Boolean operators in $\mathcal{L}$ that directly express the truth-value functions *F, N, K*, respectively. It will be enough to prove that ¬ is not classical in L; for by Theorem 3.5.2 all Boolean operators of $\mathcal{L}$ are congruential in L, hence in particular ¬ is.

Let A be any formula. Clearly,

(1) $\quad\quad\quad\quad\quad\quad\quad\quad \triangle\bot \sim \triangle(A \wedge \neg A).$

Suppose, by way of *contradictio*, that it were the case that

(2) $\quad\quad\quad\quad\quad\quad\quad\quad \neg\triangle\bot \dashv\vdash \neg\triangle(A \wedge \neg A).$

Then, by Table 3 (p. 96), $\triangle\bot \dashv\vdash \triangle(A \wedge \neg A)$. Since L is the smallest pre-classical logic in $\mathcal{L}$, it would follow, by Corollary 4.6.2, that $\triangle\bot \dashv\vDash \triangle(A \wedge \neg A)$. But both $\triangle\bot$ and $\triangle(A \wedge \neg A)$ are Boolean atoms, and hence neither tautologically implies the other. Consequently, (2) is false. Therefore, in view of (1), $\neg$ is not classical in L. ∎

The last two theorems show that a classical operator need not be congruential in pre-classical logic, nor need a congruential one be classical. These negative results should be compared with the following positive one.

THEOREM 5.3.4. *Let L be any Boolean logic. If all operators are congruential in L, then all operators are also classical in L.*

Proof. Suppose that $A \sim B$. We wish to prove that $A \dashv\vdash B$. This we do by induction on A.

By the definition of $\sim$, there are expressions X, Y and formulas C, $\overline{C}$ such that $A = X \star C \star Y$, and $B = X \star \overline{C} \star Y$, where $C \dashv\vDash \overline{C}$.

Basic step. Suppose that A is a propositional letter. Then $X = Y = \emptyset$, and so $A = C$ and $B = \overline{C}$. Thus $A \dashv\vDash B$. But the implicit matrix respects L, so in this case $A \dashv\vdash B$.

Inductive step. $A = \oplus[D_0, \ldots, D_{n-1}]$, for some n-ary operator $\oplus$ and some formulas $D_0, \ldots, D_{n-1}$ for which the claim to be proved is assumed to hold. That is to say, as our induction hypothesis we assume that, for all $i < n$ and for all formulas E, if $D_i \sim E$, then $D_i \dashv\vdash E$.

By Theorem 1.4.3, C is a subformula of A. Hence (see section 1.4) there are formulas $F_0, \ldots, F_q$, for some $q \in \omega$, such that (i) F_i is an immediate subformula of F_{i+1}, for such $i < q$; (ii) $F_0 = C$; (iii) $F_q = A$. Because of $\oplus$, $A \neq C$, so $q > 0$. Hence F_{q-1} exists, and so $F_{q-1} = D_k$, for some $k < n$. Consequently there is some $k < n$ and some formulas $D'_0, \ldots, D'_{n-1}$ such that

$D_j = D'_j$, for each $j < n$ such that $j \neq k$,

$D_k \sim D'_k$.

By assumption, $D_k \dashv\vdash D'_k$. Trivially, then, $D_i \dashv\vdash D'_i$, for all $i < n$. But $\oplus$ is congruential also by assumption. Therefore, $\oplus[D_0, \ldots, D_{n-1}] \dashv\vdash \oplus[D'_0, \ldots, D'_{n-1}]$. ∎

The difference between Theorems 3.5.3 and 3.5.4 should be carefully noted. There is of course no conflict between them, for in the proof of Theorem 5.3.3

CLASSICAL LOGICS 113

△ is not congruential, and so Theorem 5.3.4. does not apply.
The following result is straightforward but still worth listing.

THEOREM 5.3.5. *Let L be a Boolean logic. Then L is closed under* (RTE) *if and only if every operator in the language is classical in L.*

Proof. **Only-if-part.** Follows from Lemma 5.1.3.
If-part. It is enough to prove this statement: For all B, if A ~ B, then A ⊣⊢ B. This can be done by induction on the complexity of A. ∎

5.4. The lattice of classical logics

We have previously advertised that the family of classical logics would turn out to be a subfamily of the family of pre-classical ones. Now we can say which: a pre-classical logic is defined as *classical* if and only if all its operators, Boolean and non-Boolean, are classical. By Theorem 5.3.5, this means that a logic is classical if and only if it is pre-classical and closed under replacement of tautological equivalents.

In a sense this is the high point of the entire essay: we have finally arrived at the definition of classical logic. But our long labour is not yet finished — we must show (or procure some evidence for the claim) that this is the definition we want. This is the concern that will occupy us in the remainder of this chapter.

LEMMA 5.4.1. *Let L be any Boolean logic, and let $L^\$$ be the closure of L under replacement of tautological equivalents. Then $L^\$$ is Boolean, and if L is pre-classical, then so is $L^\$$.*

Proof. Define $L^\$$ as the set of all pairs $\langle \Gamma, \Theta \rangle$ such that, for some sets Γ° and Θ°, $\langle \Gamma^\circ, \Theta^\circ \rangle \in L$ and both $\Gamma \approx \Gamma^\circ$ and $\Theta \approx \Theta^\circ$. Clearly $L^\$$ is the closure of L under (RTE). To check that $L^\$$ has the right properties should be routine by now. Here we only give the least trivial case, as so often the one involving substitution.

Thus assume that L is Boolean. Suppose that $\langle \Gamma, \Theta \rangle \in L$, and let s be any substitution function. We wish to prove that $\langle s\Gamma, s\Theta \rangle \in L$.

From our assumption it follows that there are sets $\Gamma^\circ, \Theta^\circ$ such that $\Gamma \approx \Gamma^\circ$ and $\Theta \approx \Theta^\circ$, while $\langle \Gamma^\circ, \Theta^\circ \rangle \in L$. Lemma 5.2.1 implies that there are functions

$$f_1 : \Gamma \rightarrow \Gamma^\circ, \text{ and } g_1 : \Gamma^\circ \rightarrow \Gamma;$$
$$f_2 : \Theta \rightarrow \Theta^\circ, \text{ and } g_2 : \Theta^\circ \rightarrow \Theta,$$

such that

$$\forall A \in \Gamma \, (A \sim^* f_1 A), \text{ and } \forall B \in \Gamma^\circ \, (B \sim^* g_1 B);$$
$$\forall A \in \Theta \, (A \sim^* f_2 A), \text{ and } \forall B \in \Theta^\circ \, (B \sim^* g_2 B).$$

Define new functions

$$\hat{f}_1 : s\Gamma \longrightarrow s\Gamma^\circ, \text{ and } \hat{g}_1 : s\Gamma^\circ \longrightarrow s\Gamma;$$
$$\hat{f}_2 : s\Theta \longrightarrow s\Theta^\circ, \text{ and } \hat{g}_2 : s\Theta^\circ \longrightarrow s\Theta^\circ$$

as follows:

$$\forall A \in \Gamma(\hat{f}_1(sA) = s(f_1 A)), \text{ and } \forall B \in \Gamma^\circ(\hat{g}_1(sB) = s(g_1 A));$$
$$\forall A \in \Theta(\hat{f}_2(sA) = s(f_2 A)), \text{ and } \forall B \in \Theta^\circ(\hat{g}_2(sB) = s(g_2 B)).$$

Since L is Boolean, it is closed under substitution, so $\langle \Gamma^\circ, \Theta^\circ \rangle \in L$ implies that $\langle s\Gamma^\circ, s\Theta^\circ \rangle \in L$. Hence it would suffice to establish that

$$\forall A \in \Gamma(sA \sim^* sf_1 A), \text{ and } \forall B \in \Gamma^\circ(sB \sim^* sg_1 B);$$
$$\forall A \in \Theta(sA \sim^* sf_2 A), \text{ and } \forall B \in \Theta^\circ(sB \sim^* sg_2 B).$$

But this follows at once from what we have and Lemma 5.1.2. ∎

Let $\mathcal{L}$ be a language and **M** a truth-value functionally complete matrix for $\mathcal{L}$. Let X be the set of all classical logics in $\mathcal{L}$ that are respected by **M**. The inconsistent logic in $\mathcal{L}$ is respected by any matrix for $\mathcal{L}$, so X is always non-empty. Note that whenever Y is a non-empty subset of X, then $\bigcap Y$ is a member of X.

Two members of X stand out. One we just mentioned: the inconsistent logic, which we shall denote by $\mathbf{1}_{\mathcal{L},\mathbf{M}}$ or just **1**. (For $\mathbf{1}_{\mathcal{L},\mathbf{M}}$ does not really depend on **M**!) The other outstanding member of X is the smallest classical logic, that is, the intersection of all classical logics in $\mathcal{L}$ respected by **M**. We shall denote it by $\mathbf{0}_{\mathcal{L},\mathbf{M}}$, when it is desirable to emphasize its dependence on $\mathcal{L}$ and **M**, and otherwise by **0**. Thus

$$\mathbf{0} = \bigcap X.$$

It is trivial that **1** is congruential. That also **0** is congruential is more interesting.

THEOREM 5.4.2. **0** *is congruential.*

Proof. Let us denote by $L^\$$ the closure under (RTE) of the set

$$\{\langle \Gamma, \Theta \rangle : \Gamma \vDash^\mathbf{M} \Theta\}.$$

By Theorem 3.2.1 and Lemma 5.4.1, $L^\$$ is a pre-classical logic. Closed under (RTE), $L^\$$ is even classical. And no classical logic respected by **M** could be smaller than $L^\$$. Hence $\mathbf{0} = L^\$$.

Equipped with this analysis of **0** we can now tackle the problem of congruentiality. Let $\oplus$ be any n-ary operator. (By Theorem 3.5.2 it would be enough to treat only the case when $\oplus$ is non-Boolean. However, the argument that follows goes through for all operators.) Assume that $A_i \dashv\vdash B_i$, for all $i < n$. In order to establish that $\oplus$ is congruential we must show that $\oplus[A_0,\ldots, A_{n-1}] \dashv\vdash \oplus[B_0,\ldots, B_{n-1}]$. Now, since $\mathbf{0} = L^\$$, our assumption implies that, for each $i < n$, there are C_i and D_i such that $A_i \sim^* C_i$, $C_i \dashv\vDash D_i$, $D_i \sim^* B_i$. Hence $A_i \sim^* B_i$, for all $i < n$. By Lemma 5.1.3, therefore,

$$\oplus[A_0,\ldots, A_{n-1}] \sim^* \oplus[B_0,\ldots, B_{n-1}].$$

But $\mathbf{0}$ is classical and so satisfies condition (RTE). Hence

$$\oplus[A_0,\ldots, A_{n-1}] \dashv\vdash \oplus[B_0,\ldots, B_{n-1}]. \blacksquare$$

By *the lattice of classical logics in $\mathcal{L}$ with respect to* **M** we shall mean the triple $\langle X, \frown, \smile \rangle$, where, for all $L_1, L_2 \in X$,

$$L_1 \frown L_2 = L_1 \cap L_2,$$

$$L_1 \smile L_2 = \bigcap \{L \in X : L_1 \subseteq L \text{ and } L_2 \subseteq L\}.$$

This lattice we denote variously by $\mathbf{L}_{\mathcal{L},\mathbf{M}}$ and $\mathbf{L}$. It is readily seen that the definitions of the lattice operations $\frown$ ('meet') and $\smile$ ('join') (not to be confused with the set theoretical $\cap$ and $\cup$!) are meaningful: it was with an eye to the definition of $\smile$ that we admitted $\mathbf{1}$ as a member of X. Moreover, $\frown$ and $\smile$ are closed on X in the sense that if L_1 and L_2 are elements of X, then so are $L_1 \frown L_2$ and $L_1 \smile L_2$. (For the elements of lattice theory, see for example Rasiowa and Sikorski (1963).)

What we should like to prove is this: *The structure of $\mathbf{L}_{\mathcal{L},\mathbf{M}}$ does not depend on what Boolean operators happen to be primitive in $\mathcal{L}$.* It will take some effort to arrive at that result.

5.5. The Isomorphism Theorem

Fix any two languages $\mathcal{L}_1$ and $\mathcal{L}_2$ that differ at most in what primitive Boolean operators they have. Thus $\mathcal{L}_1$ and $\mathcal{L}_2$ have the same propositional letters and the same non-Boolean operators. Formally we may write

$$\mathcal{L}_1 = \langle \text{Lett}, \text{Bop}_1, \text{Iop}, R_1 \rangle, \text{ and } \mathcal{L}_2 = \langle \text{Lett}, \text{Bop}_2, \text{Iop}, R_2 \rangle,$$

where the rank functions R_1 and R_2 agree where they are both defined. Furthermore, assume that there are matrices $\mathbf{M}_1$ for $\mathcal{L}_1$ and $\mathbf{M}_2$ for $\mathcal{L}_2$ such that there are classical logics in $\mathcal{L}_1$ respected by $\mathbf{M}_1$ and classical logics in $\mathcal{L}_2$ respected by $\mathbf{M}_2$. Let $\mathbf{L}_1$ and $\mathbf{L}_2$ denote the lattices of classical logics in $\mathcal{L}_1$ with respect to $\mathbf{M}_1$ and $\mathcal{L}_2$ with respect to $\mathbf{M}_2$, respectively. Let us write

$$\mathbf{L}_1 = \langle X_1, \frown, \smile \rangle, \text{ and } \mathbf{L}_2 = \langle X_2, \frown, \smile \rangle.$$

The result that we should like to obtain can now be stated quite succinctly: we should like to prove that $\mathbf{L}_1$ and $\mathbf{L}_2$ are isomorphic. That is to say, we should like to find a function $\Phi : X_1 \to X_2$ that can be shown to be a lattice isomorphism. It would be enough to show that (i) Φ is one-to-one; (ii) Φ is onto; and (iii) Φ is order-preserving in the sense that, for all $L, L' \in X_1$, we have $L \subseteq L'$ iff $\Phi L \subseteq \Phi L'$.

Let I, J, K be the initial segments of ω (not necessarily proper!) such that $Bop_1 = \{①_i : i \in I\}$, $Bop_2 = \{②_j : j \in J\}$, $Iop = \{\triangle_k : k \in K\}$. (Concerning segments of ω: see section 1.2.) Let us denote the rank of $①_i$ by r_i, the rank of $②_j$ by s_j, and the rank of $\triangle_k$ by t_k (where i, j, k are members of I, J, K, respectively). Due to the agreement between R_1 and R_2, all this is meaningful.

Since $\mathbf{0}_1 = \mathbf{0}_{\mathcal{L}_1, \mathbf{M}_1}$ is classical and *a fortiori* pre-classical, any truth-value function directly expressed under $\mathbf{M}_2$ by a primitive Boolean operator of $\mathcal{L}_2$ is expressible also in $\mathbf{0}_1$. This means that, for each $j \in J$, there is some formula X_j of $\mathcal{L}_1$ such that X_j expresses $\mathbf{M}_2(②_j)$ with respect to some distinct propositional letters $P_0, \ldots, P_{s_j-1}$. Hence, for every assignment f of truth-values for the set of Boolean atoms in $\mathcal{L}_1$,

(#) $\qquad f^{\mathbf{M}_1}(X_j) = \mathbf{M}_2(②_j)(fP_0, \ldots, fP_{s_j-1})$.

By the same token, since $\mathbf{0}_2 = \mathbf{0}_{\mathcal{L}_2, \mathbf{M}_2}$ is classical, there is some formula Y_i of $\mathcal{L}_2$, for each $i \in I$, such that Y_i expresses $\mathbf{M}_1(①_i)$ with respect to some distinct propositional letters $Q_0, \ldots, Q_{r_i-1}$; and so, for each assignment f of truth-values for the set of Boolean atoms in $\mathcal{L}_2$,

(♭) $\qquad f^{\mathbf{M}_2}(Y_i) = \mathbf{M}_1(①_i)(fQ_0, \ldots, fQ_{r_i-1})$.

Let $Form_1$ and $Form_2$ be the sets of formulas of $\mathcal{L}_1$ and $\mathcal{L}_2$, respectively. We define two functions,

$$\overrightarrow{} : Form_1 \to Form_2, \text{ and } \overleftarrow{} : Form_2 \to Form_1,$$

as follows:

$$\overrightarrow{P} = P, \text{ if P is a propositional letter};$$
$$\overrightarrow{①_i[A_0, \ldots, A_{r_i-1}]} = Y_i(Q_0/\overrightarrow{A_0}, \ldots, Q_{r_i-1}/\overrightarrow{A_{r_i-1}}),$$
$$\overrightarrow{\triangle_k[A_0, \ldots, A_{t_k-1}]} = \triangle_k[\overrightarrow{A_0}, \ldots, \overrightarrow{A_{t_k-1}}],$$

for all $i \in I$ and $k \in K$.

$$\overleftarrow{P} = P, \text{ if P is a propositional letter},$$
$$\overleftarrow{②_j[B_0, \ldots, B_{s_j-1}]} = X_j(P_0/\overleftarrow{B_0}, \ldots, P_{s_j-1}/\overleftarrow{B_{s_j-1}}),$$
$$\overleftarrow{\triangle_k[B_0, \ldots, B_{t_k-1}]} = \triangle_k[\overleftarrow{B_0}, \ldots, \overleftarrow{B_{t_k-1}}],$$

for all $j \in J$ and $k \in K$. Thanks to the fact that $\mathcal{L}_1$ and $\mathcal{L}_2$ share propositional letters and non-Boolean operators, the definitions of $\overrightarrow{}$ and $\overleftarrow{}$ are meaningful.

If s is any substitution function in $\mathcal{L}_2$ we define a substitution function $\overleftarrow{s}$ in $\mathcal{L}_1$ as follows: for P a propositional letter, $\overleftarrow{s}P = \overleftarrow{sP}$. The definition is meaningful, as $\mathcal{L}_1$ and $\mathcal{L}_2$ have exactly the same propositional letters.

CLASSICAL LOGICS

LEMMA 5.5.1A. *For all formulas A in $\mathcal{L}_2$, $\overleftarrow{s}\overleftarrow{A} = \overleftarrow{sA}$.*

Proof. The proof is by induction. If A is a propositional letter, the lemma holds by definition. Suppose $A = ②_j[B_0,\ldots, B_{s_j-1}]$ and (the induction hypothesis) that the lemma holds for $B_0,\ldots, B_{s_j-1}$. Then

$$\overleftarrow{s}\overleftarrow{②_j[B_0,\ldots, B_{s_j}]} = \overleftarrow{s}X_j(P_0/\overleftarrow{B_0},\ldots, P_{s_j}/\overleftarrow{B_{s_j}}), \text{ by definition of } \overleftarrow{},$$

$$= X_j(P_0/\overleftarrow{s}\overleftarrow{B_0},\ldots, P_{s_j-1}/\overleftarrow{s}\overleftarrow{B_{s_j-1}}), \text{ by the definition of substitution,}$$

$$= X_j(P_0/\overleftarrow{sB_0},\ldots, P_{s_j-1}/\overleftarrow{sB_{s_j-1}}), \text{ by the induction hypothesis,}$$

$$= \overleftarrow{②_j[sB_0,\ldots, sB_{s_j-1}]}, \text{ by the definition of } \overleftarrow{},$$

$$= \overleftarrow{s②_j[B_0,\ldots, B_{s_j-1}]}, \text{ by the definition of substitution.}$$

Finally, suppose $A = \triangle_k[B_0,\ldots, B_{t_k-1}]$ and (the induction hypothesis) that the lemma holds for $B_0,\ldots, B_{t_k-1}$. Then, similarly,

$$\overleftarrow{s}\overleftarrow{\triangle_k[B_0,\ldots, B_{t_k-1}]} = \overleftarrow{s}\triangle_k[\overleftarrow{B_0},\ldots, \overleftarrow{B_{t_k-1}}]$$

$$= \triangle_k[\overleftarrow{s}\overleftarrow{B_0},\ldots, \overleftarrow{s}\overleftarrow{B_{t_k-1}}]$$

$$= \triangle_k[\overleftarrow{sB_0},\ldots, \overleftarrow{sB_{t_k-1}}]$$

$$= \overleftarrow{\triangle_k[sB_0,\ldots, sB_{t_k-1}]}$$

$$= \overleftarrow{s\triangle_k[B_0,\ldots, B_{t_k-1}]}.$$

This finishes the proof of the lemma. ∎

COROLLARY 5.5.2A. *If $\Sigma \subseteq \text{Form}_2$, then $\overleftarrow{s}\overleftarrow{\Sigma} = \overleftarrow{s\Sigma}$.*

Proof. Remark that

$$A \in \overleftarrow{s}\overleftarrow{\Sigma} \text{ iff } A = \overleftarrow{s}\overleftarrow{B}, \text{ for some } B \in \Sigma,$$

$$\text{iff } A = \overleftarrow{sB}, \text{ for some } B \in \Sigma,$$

$$\text{iff } A \in \overleftarrow{s\Sigma}. \blacksquare$$

We can of course define a dual notion $\overrightarrow{s}$, meaningful if s is a substitution function in $\mathcal{L}_1$, and prove similar results for it. That is to say, if s is any substitution function in $\mathcal{L}_1$, then the stipulation $\overrightarrow{s}P = \overrightarrow{sP}$, for P a propositional letter, defines a substitution function $\overrightarrow{s}$ in $\mathcal{L}_2$. And:

LEMMA 5.5.1B. *For all formulas A in $\mathcal{L}_1$, $\overrightarrow{s}\overrightarrow{A} = \overrightarrow{sA}$.*

COROLLARY 5.5.2B. *If $\Sigma \subseteq \text{Form}_1$, then $\overrightarrow{s}\overrightarrow{\Sigma} = \overrightarrow{s\Sigma}$.*

LEMMA 5.5.3. *Let f be any truth-value assignment for the set of propositional* letters. Suppose that A and B are purely Boolean formulas of $\mathcal{L}_1$ and $\mathcal{L}_2$ respectively. Then $f^{M_1}(A) = f^{M_2}(\vec{A})$, and $g^{M_2}(B) = g^{M_1}(\vec{B})$.

Proof. 'We prove the former claim; the proof of the latter one is analogous. The proof is inductive. If A is a propositional letter, there is nothing to prove. If A is not a propositional letter, then, since A contains only Boolean operators, $A = \mathcal{O}_i[C_0, \ldots, C_{r_i-1}]$, for some i and some $C_0, \ldots, C_{r_i-1}$. Assume that the lemma holds for $C_0, \ldots, C_{r_i-1}$. Then

$$f^{M_1}(\mathcal{O}_i[C_0, \ldots, C_{r_i-1}]) = M_1(\mathcal{O}_i)(f^{M_1}C_0, \ldots, f^{M_1}C_{r_i-1})$$
$$= M_1(\mathcal{O}_i)(f^{M_2}\vec{C_0}, \ldots, f^{M_2}\vec{C_{r_i-1}}).$$
$$= f^{M_2}(Y_i(Q_0/\vec{C_0}, \ldots, Q_{r_i-1}/\vec{C_{r_i-1}}))$$
$$= f^{M_2}\overrightarrow{(\mathcal{O}_i[C_0, \ldots, C_{r_i-1}])}. \blacksquare$$

LEMMA 5.5.4A. *If A and B are any formulas in $\mathcal{L}_2$, then $A \models^{M_2} B$ iff $\overleftarrow{A} \models^{M_1} \overleftarrow{B}$.*

Proof. A and B may contain non-Boolean operators. However — cf. Lemma 3.1.1. — we can always find A_0 and B_0 and a substitution function s such that the following conditions hold:

(1) A_0 and B_0 are purely Boolean.
(2) s is one-to-one.
(3) sP is a Boolean atom, for every propositional letter P.
(4) $A = sA_0$.
(5) $B = sB_0$.

Note that (2) and (3) imply that

(6) $\overleftarrow{s}$ is one-to-one;
(7) $\overleftarrow{s}P$ is a Boolean atom, for every propositional letter P.

The proof of the lemma is now straightforward:

$A \models^{M_2} B$ iff $sA_0 \models^{M_2} sB_0$, by (4) and (5),

iff $A_0 \models^{M_2} B_0$, because of (2) and (3),

iff $\overleftarrow{A_0} \models^{M_1} \overleftarrow{B_0}$, by (1) and Lemma 5.5.3,

iff $\overleftarrow{s}\overleftarrow{A_0} \models^{M_1} \overleftarrow{s}\overleftarrow{B_0}$, because of (6) and (7),

iff $\overleftarrow{sA_0} \models^{M_1} \overleftarrow{sB_0}$, by Lemma 5.5.1A,

iff $\overleftarrow{A} \models^{M_1} \overleftarrow{B}$, by (4) and (5). $\blacksquare$

COROLLARY 5.5.5A. *If A and B are formulas in $\mathcal{L}_2$, then $A \dashv\models^{M_2} B$ iff $\overleftarrow{A} \dashv\models^{M_1} \overleftarrow{B}$.*

CLASSICAL LOGICS

In the same way the 'duals' of Lemma 5.5.4A and Corollary 5.5.5A may be derived:

LEMMA 5.5.4B. *If A and B are formulas in $\mathcal{L}_1$, then* $A \vDash^{M_1} B$ *iff* $\overrightarrow{A} \vDash^{M_2} \overrightarrow{B}$.

COROLLARY 5.5.5B. *If A and B are formulas in $\mathcal{L}_1$, then* $A \dashv\vDash^{M_1} B$ *iff* $\overrightarrow{A} \dashv\vDash^{M_2} \overrightarrow{B}$.

In the long build-up towards the Isomorphism Theorem 5.5.10, the next result is where the question of classicalness arises.

THEOREM 5.5.6. *Let L be any logic in X_1. Then, for all formulas A in $\mathcal{L}_1$, $A \dashv\vdash_L \overleftrightarrow{A}$. Similarly, if L is any logic in X_2, then, for all formulas B in $\mathcal{L}_2$, $B \dashv\vdash_L \overleftrightarrow{B}$.*

Proof. The two claims of the theorem are proved separately (though analogously). We shall give the proof of the former claim only. Since L is classical, it will be enough to prove that, for all A in $\mathcal{L}_1$,

(†) $$A \sim^* \overleftrightarrow{A}.$$

The proof is inductive on A. The basic step is trivial: if A is a propositional letter, then the claim (†) obviously holds.

The inductive step consists of two parts. Assume first that $A = \triangle_k[B_0,\ldots, B_{t_k-1}]$, and that for each B_m, where $m < t_k$, the claim holds. This is the easy part, for

$$\overleftrightarrow{\triangle_k[B_0,\ldots, B_{t_k-1}]} = \triangle_k[\overleftrightarrow{B_0},\ldots, \overleftrightarrow{B_{t_k-1}}],$$

and so we may just appeal to Lemma 5.1.3.

Assume now instead that $A = \textcircled{1}_i[B_0,\ldots, B_{r_i-1}]$, for some i, and that for each B_m, where $m < r_i$, the claim holds. For this part a more elaborate argument is needed.

Let us first recall that there is a formula Y_i of $\mathcal{L}_2$ that expresses $M_1(\textcircled{1}_i)$ with respect to some propositional letters $Q_0,\ldots, Q_{r_i-1}$, and that

$$\overrightarrow{\textcircled{1}_i[Q_0,\ldots, Q_{r_i-1}]} = Y_i,$$

$$\overrightarrow{\textcircled{1}_i[B_0,\ldots, B_{r_i-1}]} = Y_i(Q_0/\overrightarrow{B_0},\ldots, Q_{r_i-1}/\overrightarrow{B_{r_i-1}}).$$

Let Z be the formula of $\mathcal{L}_1$ such that $Z = \overleftarrow{Y_i}$. Then

(1) $$\overleftrightarrow{\textcircled{1}_i[B_0,\ldots, B_{r_i-1}]} = Z(Q_0/\overleftrightarrow{B_0},\ldots, Q_{r_i-1}/\overleftrightarrow{B_{r_i-1}}).$$

Let f be any truth-value assignment for the set of propositional letters. Then

$$f^{M_1}(Z) = f^{M_1}(\overleftarrow{Y_i})$$
$$= f^{M_2}(Y_i), \text{ by Lemma 5.5.3,}$$
$$= \mathbf{M}_1(\textcircled{i})(fQ_0,\ldots, fQ_{r_i-1}), \text{ by } (\flat) \text{ on p. 116,}$$
$$= f^{M_1}(\textcircled{i}[Q_0,\ldots, Q_{r_i-1}]).$$

Therefore $\textcircled{i}[Q_0,\ldots, Q_{r_i-1}] \dashv\vdash^{M_1} Z$, and so certainly $\textcircled{i}[Q_0,\ldots, Q_{r_i-1}] \sim^* Z$. Hence also

(2) $\qquad \textcircled{i}[B_0,\ldots, B_{r_i-1}] \sim^* Z(Q_0/B_0,\ldots, Q_{r_i-1}/B_{r_i-1}).$

By Lemma 5.1.3, and the induction hypothesis,

(3) $\qquad Z(Q_0/B_0,\ldots, Q_{r_i-1}/B_{r_i-1}) \sim^* Z(Q_0/\overleftarrow{B_0},\ldots, Q_{r_i-1}/\overrightarrow{B_{r_i-1}}).$

Consequently, by (1), (2) and (3),

$$\textcircled{i}[B_0,\ldots, B_{r_i-1}] \sim^* \overleftarrow{\textcircled{i}[B_0,\ldots, \overrightarrow{B_{r_i-1}}]},$$

as we wanted to show. ∎

COROLLARY 5.5.7. *If L is a classical logic in* $\mathbf{L}_1$ *and* Γ *and* Θ *are sets of formulas in* $\mathcal{L}_1$, *then* $\Gamma \vdash_L \Theta$ *iff* $\overleftarrow{\overrightarrow{\Gamma}} \vdash_L \overleftarrow{\overrightarrow{\Theta}}$. *Similarly, if L is a classical logic in* $\mathbf{L}_2$ *and* Δ *and* Λ *are sets of formulas in* $\mathcal{L}_2$, *then* $\Delta \vdash_L \Lambda$ *iff* $\overrightarrow{\overleftarrow{\Delta}} \vdash_L \overrightarrow{\overleftarrow{\Lambda}}$.

Proof. We prove half of the first part. Assume that

(1) $\qquad\qquad\qquad \Gamma \vdash \Theta.$

By Theorem 5.5.6, $\overrightarrow{\overleftarrow{A}} \vdash A$, for all A (in $\mathcal{L}_1$). Hence, by monotonicity,

(2) $\qquad\qquad\qquad \overrightarrow{\overleftarrow{\Gamma}} \vdash A, \text{ for all } A \in \Gamma.$

By (1), (2), and cut,

(3) $\qquad\qquad\qquad \overrightarrow{\overleftarrow{\Gamma}} \vdash \Theta.$

But by Theorem 5.5.6, also $B \vdash \overleftarrow{\overrightarrow{B}}$, for all B. Hence, by monotonicity,

(4) $\qquad\qquad\qquad B \vdash \overleftarrow{\overrightarrow{\Theta}}, \text{ for all } B \in \Theta.$

By (3), (4), and cut, $\overleftarrow{\overrightarrow{\Gamma}} \vdash \overleftarrow{\overrightarrow{\Theta}}$. ∎

LEMMA 5.5.8A. *If L is a classical logic in $\mathbf{L}_1$, then the set*

$$\Phi L = \{\langle \Delta, \Lambda \rangle : \Delta, \Lambda \subseteq \text{Form}_2 \ \& \ \langle \overleftarrow{\Delta}, \overleftarrow{\Lambda} \rangle \in L\}$$

is a classical logic in $\mathbf{L}_2$.

Proof. Suppose that $L \in X_1$. We must show that $\Phi L \in X_2$. That is to say, we must show that ΦL is a classical logic in $\mathcal{L}_2$ respected by $\mathbf{M}_2$. So we must go through the usual list of conditions and check that they are all satisfied. We shall be content to give some examples, leaving the others (that are simpler) to the reader.

Substitution. Assume that Δ and Λ are any sets of formulas in $\mathcal{L}_2$ such that $\langle \Gamma, \Lambda \rangle \in \Phi L$. Let s be any substitution function in $\mathcal{L}_2$. In order to prove that ΦL is closed under substitution, we must show that also $\langle s\Delta, s\Lambda \rangle \in \Phi L$. But this is straightforward:

$\langle \Delta, \Lambda \rangle \in \Phi L$ implies $\langle \overleftarrow{\Delta}, \overleftarrow{\Lambda} \rangle \in L$, by definition of Φ,

implies $\langle \overleftarrow{s}\overleftarrow{\Delta}, \overleftarrow{s}\overleftarrow{\Lambda} \rangle \in L$, since L is closed under substitution,

implies $\langle \overleftarrow{s\Delta}, \overleftarrow{s\Lambda} \rangle \in L$, by Corollary 5.5.2A,

implies $\langle s\Delta, s\Lambda \rangle \in \Phi L$, by definition of Φ.

Type determination. Next we show that every Boolean operator of $\mathcal{L}_2$ is type determined in ΦL. Take any Boolean operator $②_j$ and any partitioning $\langle I, J \rangle$ of s_j. Suppose that $x_0, \ldots, x_{s_j-1}$ are truth-values such that

$$I = \{i < s_j : x_i = 1\}, \text{ and } J = \{i < s_j : x_i = 0\},$$

and

(§) $\mathbf{M}_2(②_j)(x_0, \ldots, x_{s_j-1}) = 0.$

We wish to prove that, if $P_0, \ldots, P_{s_j-1}$ are distinct propositional letters as on p. 116, then

$$②_j[P_0, \ldots, P_{s_j-1}], \{P_i : i \in I\} \vdash_{\Phi L} \{P_i : i \in J\}.$$

By definition of Φ, this holds if and only if

$$\overleftarrow{②_j[P_0, \ldots, P_{s_j-1}]}, \{P_i : i \in I\} \vdash_L \{P_i : i \in J\}.$$

By definition of $\overleftarrow{}$, this in its turn holds if and only if

$$X_i, \{P_i : i \in I\} \vdash_L \{P_i : i \in J\}.$$

Hence it will be enough to prove that

$$X_i, \{P_i : i \in I\} \vDash^{\mathbf{M}_1} \{P_i : i \in J\}.$$

Take any truth-value assignment f for the set of Boolean atoms in $\mathcal{L}_1$ such that

$$f(P_i) = 1, \text{ if } i \in I$$
$$= 0, \text{ if } i \in J.$$

It will now be enough to prove that $f^{M_1}(X_i) = 0$. But this is obvious in view of (#) on p. 116 and (§). Hence, ②$_j$ is of type 0 with respect to $\langle I, J \rangle$ in ΦL.

If the assumption (§) is replaced by

(§§) $\qquad\qquad M_2(②_j)(x_0,\ldots, x_{s_j-1}) = 1,$

then we may prove, in the same way, that ②$_j$ is of type 1 with respect to $\langle I, J \rangle$ in ΦL.

Truth-value functional completeness. Let ϕ be any truth-value function, say n-ary. Since L is truth-value functionally complete, there is some formula A in $\mathcal{L}_1$ such that A expresses ϕ in L with respect to some $P_0,\ldots, P_{n-1}$. That is to say, for every assignment f of truth-values to the propositional letters,

$$f^{M_1}(A) = \phi(fP_0,\ldots, fP_{n-1}).$$

It is now easy to see that $\vec{A}$ expresses ϕ in ΦL with respect to $P_0,\ldots, P_{n-1}$. For take any assignment g of truth-values to the propositional letters. Then, since A does not contain any non-Boolean operators, it follows from Lemma 5.5.3 that $g^{M_2}(\vec{A}) = g^{M_1}(A) = \phi(gP_0,\ldots, gP_{n-1})$. Hence, L is truth-value functionally complete.

Replacement of tautological equivalents. Suppose that A and B are formulas in $\mathcal{L}_2$ such that

(1) $\qquad\qquad A \underset{M_2}{\sim} B.$

Then there are formulas C, C' in $\mathcal{L}_2$ and expressions U_2, V_2 in $\mathcal{L}_2$ such that

(2) $\qquad\qquad A = U_2' \star C \star V_2;$

(3) $\qquad\qquad B = U_2 \star C' \star V_2;$

(4) $\qquad\qquad C \dashv\vdash^{M_2} C'.$

Let Q be a propositional letter that does not occur in U_2 or V_2. By Lemma 1.4.4, $U_2 \star Q \star V_2$ is a formula in $\mathcal{L}_2$. Hence $\overleftarrow{U_2 \star Q \star V_2}$ is a formula in $\mathcal{L}_1$. That is to say, there are uniquely defined expressions U_1 and V_1 in $\mathcal{L}_1$ such that $\overleftarrow{U_2 \star Q \star V_2} = U_1 \star Q \star V_1$. Let s be any substitution function such that, for every propositional letter $P \neq Q$, $sP = P$. It follows that $\overleftarrow{s}$ is a substitution function in $\mathcal{L}_2$ such that, for every propositional letter $P \neq Q$, $\overleftarrow{s}P = P$. Using Theorem 1.5.2 and Lemma 5.5.1A we conclude that

$$\overleftarrow{U_2 \star sQ \star V_2} = \overleftarrow{s(U_2 \star Q \star V_2)}$$
$$= \overleftarrow{\overleftarrow{s}}(\overleftarrow{U_2 \star Q \star V_2})$$
$$= \overleftarrow{\overleftarrow{s}}(U_1 \star Q \star V_1)$$
$$= U_1 \star \overleftarrow{\overleftarrow{s}}Q \star V_1.$$

If we choose s in such a way that $sQ = C$ (whence $\overleftarrow{sQ} = \overleftarrow{C}$), this result and (2) yield

(5) $\qquad \overleftarrow{A} = U_1 \star \overleftarrow{C} \star V_1.$

On the other hand, if we choose s so that $sQ = C'$ (whence $\overleftarrow{sQ} = \overleftarrow{C'}$), the same result and (3) yield

(6) $\qquad \overleftarrow{B} = U_1 \star \overleftarrow{C'} \star V_1.$

By Lemma 5.5.4A, (4) implies that

(7) $\qquad \overleftarrow{C} \dashv \vDash^{M_1} \overleftarrow{C'}.$

By (5), (6), and (7), therefore, $A \underset{M_1}{\sim} B$. But L satisfies condition (RTE). Hence $A \dashv \vdash_L B$. By the definition, then, $A \dashv \vdash_{\Phi L} B$. This is the desired result. ∎

LEMMA 5.5.8B. *If L is a classical logic in* $\mathbf{L}_2$, *then the set*

$$\Psi L = \{\langle \Gamma, \Theta \rangle : \Gamma, \Theta \subseteq \text{Form}_1 \ \& \ \langle \vec{\Gamma}, \vec{\Theta} \rangle \in L\}$$

is a classical logic in $\mathbf{L}_1$.

Proof. Dual to that of Lemma 5.5.8A. ∎

LEMMA 5.5.9. *If* Γ *and* Θ *are sets of formulas in* $\mathcal{L}_1$, *then, for all logics* $L \in X_1$, *we have* $\langle \Gamma, \Theta \rangle \in L$ *iff* $\langle \vec{\Gamma}, \vec{\Theta} \rangle \in \Phi L$. *Similarly, if* Δ *and* Λ *are sets of formulas in* $\mathcal{L}_2$, *then, for all logics* $L \in X_2$, *we have* $\langle \Delta, \Lambda \rangle \in L$ *iff* $\langle \overleftarrow{\Delta}, \overleftarrow{\Lambda} \rangle \in \Psi L$.

Proof. Follows at once from Corollary 5.5.7. ∎

Now for the crowning achievement.

THEOREM 5.5.10. (*Isomorphism Theorem.*) $\mathbf{L}_1$ *and* $\mathbf{L}_2$ *are isomorphic.*

Proof. For every $L \in X_1$, ΦL is defined and is a member of X_2, as testified by Lemma 5.5.8A. Thus we may regard Φ as a function from X_1 to X_2. Similarly, we may regard Ψ as a function from X_2 to X_1. We will now show that Φ is a lattice isomorphism from $\mathbf{L}_1$ to $\mathbf{L}_2$. As explained on p. 116, there are three parts to the proof.

Injectivity of Φ. Assume that $L, L' \in X_1$. If there are some Γ and Θ such that $\langle \Gamma, \Theta \rangle \in L' - L$, then, by Lemma 5.5.9, $\langle \vec{\Gamma}, \vec{\Theta} \rangle \in \Phi L' - \Phi L$. Hence Φ is one-to-one.

Surjectivity of Φ. Take any $L \in X_2$. Then, by Lemma 5.5.8B, $\Psi L \in X_1$. Moreover,

$$\Phi\Psi L = \{\langle \Delta, \Lambda \rangle : \langle \overleftarrow{\Delta}, \overleftarrow{\Lambda} \rangle \in \Psi L\}, \text{ by definition of } \Phi,$$
$$= \{\langle \Delta, \Lambda \rangle : \langle \overrightarrow{\overleftarrow{\Delta}}, \overrightarrow{\overleftarrow{\Lambda}} \rangle \in L\}, \text{ by definition of } \Psi,$$
$$= \{\langle \Delta, \Lambda \rangle : \langle \Delta, \Lambda \rangle \in L\}, \text{ by Corollary 5.5.7,}$$
$$= L.$$

Hence Φ is onto X_2, for, by Lemma 5.5.8B, $\Psi L \in X_1$.

Order-preservation of Φ. Suppose $L, L' \in X_1$. If $L \subseteq L'$, then

$$\langle \Delta, \Lambda \rangle \in \Phi L \text{ implies } \langle \overleftarrow{\Delta}, \overleftarrow{\Lambda} \rangle \in L, \text{ by definition of } \Phi,$$
$$\text{implies } \langle \overleftarrow{\Delta}, \overleftarrow{\Lambda} \rangle \in L', \text{ by assumption,}$$
$$\text{implies } \langle \overrightarrow{\overleftarrow{\Delta}}, \overrightarrow{\overleftarrow{\Lambda}} \rangle \in \Phi L' \text{ by Lemma 5.5.9,}$$
$$\text{implies } \langle \Delta, \Lambda \rangle \in \Phi L', \text{ by Corollary 5.5.7,}$$

so $\Phi L \subseteq \Phi L'$. A similar argument shows that $L \subseteq L'$ if $\Phi L \subseteq \Phi L'$. ∎

6
MODAL LOGICS: A POSTSCRIPT

6.1. Two ambitions

Two different ambitions have gone into the writing of this book, one to make it a textbook (to present known material in an instructive manner), the other to make it a monograph (to present new material on a certain subject). To some extent these ambitions have been in conflict.

The textbook ambition was to give a survey of some fundamental ideas in a certain department of logic, and to do so in a reasonably general way. This ambition dominates the book up through Chapter 4. A quest for generality is of course common; in a way, it is what mathematics is about. If it is true that a deeper understanding of anything at all is possible only when key ideas are isolated and the interplay between them is exposed, then generalization is indeed necessary. But there is also an obvious disadvantage to generalization: with increased simplicity at a higher level goes growing complexity at the ground level. This disadvantage is particularly problematic in a textbook. And we have certainly paid a high price in increased surface complexity in this book.

But however awkward the burden of complexity, a high level of generality and abstraction was needed in order to realize the monograph ambition. Actually there were two versions of this ambition, one original and one reduced. The original ambition was to find a satisfactory abstract definition of the concept of modal logic, of the same generality as our definitions of the concepts of common, pre-classical, and classical logic. Like so many ambitions it was never fulfilled; hence, the reduced ambition which issued in Chapter 5 and which was to treat of one particular problem relevant to the definition of modal logic, that of 'language sensitivity'.

This Postscript has been added as an apologia for the monograph ambition which might be needed, especially in view of the terseness of Chapter 5. In a way it is not part of the main text: it deals with matters not sufficiently well researched, and so the exposition is sketchier and more tentative than that of the previous chapters. In Section 6.2 we give a historical survey of one aspect of modal logic which in the following two sections is analysed with tools developed in the book. In Section 6.5, finally, we offer an example of a stipulative definition of a concept of modal logic – unfortunately a far cry from what was originally hoped for.

6.2. A question in modal logic

No other modal logician worked in so many different object-languages as the father of the discipline, C. I. Lewis. Already his first publication in this field (1912) contains the embryo of what is termed even there 'the calculus of strict implication'. The starting point is his observation that propositions like 'Either Matilda does not love me, or I am beloved' differ from propositions like 'Either Caesar died, or the moon is made of green cheese', and he feels that non-Boolean operators are needed in order to formalize the former kind. (For an interesting commentary on this idea, see Scott (1971).) Accordingly Lewis suggests that extensional (what we call Boolean) operators have intensional counterparts; he mentions disjunction and material implication in particular. Little formal work is carried out in this programmatic paper, but more is envisaged, and Lewis especially recommends that one 'retain both extensional and intensional disjunction, symbolize them differently, and define strict implication in terms of intensional disjunction only'. One natural way to read Lewis here is as proposing that a new binary non-Boolean operator v ('intensional disjunction') be added to a truth-functionally complete set of Boolean operators, $A \, v \, B$ bearing the intuitive reading of 'either A or B' or, perhaps, 'necessarily either A or B'. If strict implication is rendered by another binary operator, $\rightarrowtail$ (the 'fish-hook'), $A \rightarrowtail B$ bearing the intuitive reading 'A strictly implies B' or 'B is deducible from A', then $A \rightarrowtail B$ is supposed to be definable in terms of v and Boolean negation:

$$A \rightarrowtail B =_{df} \neg A \, v \, B.$$

When a few years later Lewis was to present his first full account of his new kind of logic, he departed somewhat from the letter of the recommendation quoted above. His primitives in (1918) are the Boolean $\neg$ and $\wedge$ and a new intensional operator for which we will employ the unorthodox symbol $\Diamond$, where $\Diamond A$ is to be read 'it is impossible that A is true' or 'A is impossible'. (Some authors, such as Kneebone (1963), view Lewis as also having the binary = as primitive (which Lewis in fact says that he does). While $A = B$ bears the intuitive reading 'A and B are strictly equivalent', Lewis uses = very much as in this section we are using the naïve symbol '$=_{df}$'.) In these terms the intensional operators contemplated in (1912) are definable; indeed, we may even define =:

$$A \, v \, B =_{df} \Diamond(\neg A \wedge \neg B),$$

$$A \rightarrowtail B =_{df} \Diamond(A \wedge \neg B),$$

$$A = B =_{df} \Diamond(A \wedge \neg B) \wedge \Diamond(B \wedge \neg A).$$

An operator new to (1918) is the binary $\circ$, a kind of intensional conjunction: $A \circ B$ is to have the intuitive reading 'it is possible that A and B both be true' or 'A and B are consistent'. This operator, too, is definable:

$$A \circ B =_{df} \neg\Diamond(A \wedge B).$$

It is worth noting that in the same work Lewis also contemplates a Calculus of Consistencies with $\neg, \Diamond, \circ$ (and $=$) as primitive operators, as well as a Calculus of Ordinary Inference with $\neg, \wedge, \strictif$ (and $=$) as primitive. In the former calculus he adopts the definitions:

$$A \vee B =_{df} \neg(\neg A \circ \neg B),$$
$$A \strictif B =_{df} \neg(A \circ \neg B),$$

in the latter

$$A \vee B =_{df} \neg A \strictif B,$$
$$A \circ B =_{df} \neg(A \strictif \neg B),$$
$$A = B =_{df} (A \strictif B) \wedge (B \strictif A).$$

In Lewis and Langford (1932) we meet again with a new alphabet: $\neg$ and $\wedge$ remain primitive, but $\Diamond$ has been replaced by a new unary non-Boolean operator $\lozenge$ (the 'diamond'), $\lozenge A$ bearing the intuitive reading 'it is possible that A is true' or 'A is possible'. The preceding concepts may be redefined in this new context:

$$\Diamond A =_{df} \neg \lozenge A,$$
$$A \strictif B =_{df} \neg \lozenge(A \wedge \neg B),$$
$$A = B =_{df} \neg \lozenge(A \wedge \neg B) \wedge \neg \lozenge(B \wedge \neg A),$$
$$A \circ B =_{df} \neg(A \strictif \neg B).$$

Here it may be noted that what the definition of $\circ$ actually amounts to in primitive terms is $A \circ B =_{df} \neg \neg \lozenge(A \wedge \neg \neg B)$. The definition $A \circ B =_{df} \lozenge(A \wedge B)$ would of course do just as well.

Even though Lewis dealt with necessity as well as with possibility, for some reason he never introduced an operator to capture the former and match $\lozenge$. This was done instead by Gödel (1933). The customary symbol used nowadays (and evidently due to Fitch) is $\square$ (the 'box'), where $\square A$ is to be read 'it is necessary that A is true' or 'A is necessary'. (Actually, Gödel himself used the symbol B, for the German *beweisbar*, and he was interested not so much in necessity as in provability and provability in a certain formal system. For recent work related to this, see Boolos (1979).) Again all previous concepts can be defined:

$$\lozenge A =_{df} \neg \square \neg A,$$
$$A \strictif B =_{df} \square(A \rightarrow B),$$

etc.

The preceding account would seem to include all modal operators that have actually enjoyed any currency. There are others, though. Montgomery and Routley (1966) introduced two new unary operators $\triangle$ and $\triangledown$ which they regarded as philosophically important, $\triangle A$ with the intuitive reading 'it is non-contingent that A', and $\triangledown A$ with the intuitive reading 'it is contingent that A'. In a logic based on $\triangle$, the construct $\square A$ is meant to be definable as $A \wedge \triangle A$, while in a logic based on $\triangledown$, the construct $\Diamond A$ is meant to be definable as $A \vee \triangledown A$. However, these translations are of more restricted applicability than the preceding, for the proposed definitions accord with intuitive preconceptions only in logics in which $\square A \vdash A$ (or, dually, $A \vdash \Diamond A$).

Probably the strangest modal operators in the literature are those studied in Halldén (1949); they are two ternary operators which we will write $\wedge$ and $\vee$, and they are explained by the 'equations'

$$A, B \wedge C =_{df} \neg A \wedge \neg B \wedge \neg \Diamond C,$$

$$A, B \vee C =_{df} \neg A \vee \neg B \vee \neg \Diamond C.$$

Idiomatic readings of these operators are lacking. It should be noted that Halldén never thought of them as formalizing any natural concepts. Their interest is purely theoretical as a kind of modal Sheffer strokes: each by itself, unaided by Boolean operators, suffices to develop large parts of modal logic. However, as in the case of Montgomery and Routley's operators, those of Halldén only work for logics in which $\square A \vdash A$ (or, dually, $A \vdash \Diamond A$); in the case of $\wedge$ also normality is required (for a definition of normality, see Section 6.5 below).

Among the many possible operators that have never been proposed by anyone, there is one that should be mentioned here, the unary ϕ, with ϕA bearing the intuitive reading 'it is not necessary that A' or 'A is non-necessary'. The concept of non-necessity does not appear to equal in intuitive significance that of impossibility, let alone those of necessity or possibility. But from a theoretical point of view, ϕ is on a par with $\Diamond$ as well as with $\square$ and $\Diamond$.

The popularity of the many operators mentioned has varied considerably over the years. In the early days $\Diamond$ was used, and even today an entailment logician may use $\rightarrow$ as primitive in order to study certain fragments. But while the choice of Boolean operators continues to vary, it is rare nowadays to find other modal primitives than $\square$ and $\Diamond$. If to our previous code names T, F, N, K, A, C, E we add L, M, N', A', C', E' for $\square, \Diamond, \phi, \nu, \rightarrow, =$, respectively, we are able to encode the primitive operators of most actual presentations. Lewis and Langford (1932) is perhaps the most influential text in modal logic, and its N-K-M base has been correspondingly popular. Lemmon (1957) launched the Gödel type base N-C-L, since widely adopted. Kripke (1963), a paradigm paper, uses N-K-L, while the best-selling modern text-book in the field, Hughes and Cresswell (1968), is based on N-A-L. Lemmon (1977) uses F-C-L, an interesting combination, and this work has not been without impact. All the preceding authors were concerned to adopt a minimal base, but this is by no means always the case. von Wright (1951) seems to use N-K-A-C-E-L-M as his base, and so does the recent text Chellas (1980). The situation is complicated by the fact that sometimes authors do not specify what operators they regard as primitive.

Against a background of so much diversity, several questions emerge. Perhaps the most obvious one is whether results proved with respect to one object language also hold for other object languages. Some typical results in early modal logic were that S2 is decidable (McKinsey); that S4 has fourteen irreducible non-equivalent modalities (Parry); that no logic between S1 and S5 has a finite characteristic matrix (Dugundji); that any logic extending S5 is normal (Scroggs). When such results were first proved, the proofs would be carried out with respect to some object language or other. At the same time it was no doubt taken for granted that the results reached will hold no matter what 'reasonable' object language is chosen. But clearly it is an interesting problem in the foundations of modal logic to articulate presuppositions that would justify such a belief:

QUESTION. *What results of modal logic are language independent under what conditions?*

Readers familiar with the preceding two chapters will be able to fabricate counterexamples to one kind of carelessly formulated 'results' in modal logic. For example, it is *not* a general result that every formula has to be either a thesis or an antithesis of a Post complete modal logic if it is built from a set of tautologies or contradictions. But in any logic that is classical, in the sense of this book, the result holds. (Sketch of proof of the former contention: Assume that $\neg, \wedge, \vee,$ and $\square$ are primitive, and let P be a propositional letter. The set $\{\square((P \wedge P) \wedge \neg(P \wedge P)), \neg\square((P \vee P) \wedge \neg(P \vee P))\}$ is consistent in pre-classical logic, and so there is a Post complete pre-classical logic extending it. Since this logic is closed under substitution, it follows that $\square(P \wedge \neg P) \not\vdash$ and $\not\vdash \square(P \wedge \neg P)$.)

A question of less practical importance but still of philosophical interest concerns the nature of entities like S3. In C. I. Lewis' original formulation in (1918) the nowadays never-heard-of N–K–N' base was used. In Lewis and Langford (1932) it was given a reformulation on the N–K–M base. Both Lemmon (1957) and Kripke (1965) have interesting things to say about S3, but on the N–C–L, respectively, the N–K–L base. These are just four of the literally infinitely many possibilities there are. In what sense is it S3 that we are talking about in all cases? A Halldén version of S3 based on ∨ would look different from anything you have ever seen — would it still be S3?

Yet another question concerns the nature of modal logic itself: what do all or most systems of modal logic have in common? Is there anything to what we are used to calling modal logics as distinct from, say, the logics of counterfactuals as developed by Stalnaker and David Lewis?

The work done in this book does not answer these questions; however, it provides some terms in which they can be discussed, and perhaps it even throws some light on them.

6.3. Some translations between particular modal languages

In this section we will examine some particular object-languages used by modal logicians. To facilitate discussion we will make the following assumptions. First, by Theorem 5.5.10, the Boolean component of a language is not very interesting in the context of classical logic, and as we will deal almost exclusively with classical logics from now on, we may as well suppose that our languages come with some unspecified but sufficiently rich set of Boolean operators. More precisely, we will understand that there is a certain matrix associated with this set of Boolean operators which is implicit in any logic in the language, and that in fact any logic in the language is at least pre-classical. Let us introduce the term *truth-value functionally complete language* to refer to such a language. Furthermore, we shall assume that we operate with some fixed denumerable set of propositional letters: every language mentioned in this section and the next is supposed to have this set for its set of propositional letters. Neither assumption entails any real loss of generality.

Let us adopt the convention that if $\mathcal{L}_s$ is a language where s is any subscript, possibly empty, then the set of formulas in $\mathcal{L}_s$ may be denoted by Form$_s$ and the smallest classical logic in $\mathcal{L}_s$ by $\mathbf{0}_s$. If $\mathcal{L}$ is a language with $\oplus_0, \ldots, \oplus_k$ as its only non-Boolean operators, then we will sometimes find it convenient to write $\mathcal{L} = \mathcal{L}_{\oplus_0 \ldots \oplus_k}$.

We will now establish an improved version of Theorem 5.5.10:

THEOREM 6.3.1. *For* $i = 1, 2$, *let $\mathcal{L}_i$ be a truth-value functionally complete language and $\mathbf{L}_i$ the lattice of all classical logics in $\mathcal{L}_i$ that extend a certain classical logic* L_i *in $\mathcal{L}_i$. Suppose that L_1 and L_2 are syntactically equivalent with respect to some translations* f : Form$_1 \longrightarrow$ Form$_2$ *and* g : Form$_2 \longrightarrow$ Form$_1$. *Then $\mathbf{L}_1$ and $\mathbf{L}_2$ are isomorphic under the map*

$$\Phi: L \longmapsto \{\langle \Delta, \Lambda \rangle : \Delta, \Lambda \subseteq \text{Form}_2 \ \& \ \langle g\Delta, g\Lambda \rangle \in L\}.$$

Proof. Let us call Φ as defined *the map induced by* f *and* g; this term will be used not only in this proof. That $\Phi L_1 = L_2$ follows from the definition of Φ and the assumption that L_1 and L_2 are syntactically equivalent with respect to f and g. The rest of the proof consists in checking that the proof of Theorem 5.5.10 can be taken over with appropriate adjustment. It may be noted that Theorem 5.5.6 now comes for free, for from the syntactic equivalence of L_1 and L_2 we at once infer that $A \dashv\vdash gfA$ throughout $\mathbf{L}_1$, for all $A \in$ Form$_1$, and that $B \dashv\vdash fgB$ throughout $\mathbf{L}_2$, for all $B \in$ Form$_2$. ∎

As a further preliminary step we will give a semantic characterization of the smallest classical logic $\mathbf{0}$ in a truth-value functionally complete language $\mathcal{L}$. Let us say that a truth-value assignment v in $\mathcal{L}$ *respects (RTE)* if $\bar{v}(A) = \bar{v}(B)$ whenever $A \sim^* B$, where $\sim^*$ is the relation of replacement of tautological equivalents defined in Section 5.1. Furthermore, let us write $\Gamma \vDash_{\text{RTE}} \Theta$ if Γ, Θ are sets of formulas in $\mathcal{L}$ such that, for all (RTE)-respecting valuations v, if $\bar{v}(A) = 1$, for all $A \in \Gamma$, then $\bar{v}(B) = 1$, for some $B \in \Theta$.

LEMMA 6.3.2. $\mathbf{0} = \{\langle \Gamma, \Theta \rangle : \Gamma, \Theta \subseteq \text{Form} \,\&\, \Gamma \vDash_{\text{RTE}} \Theta\}$.

Proof. By the proof of Theorem 5.4.2, $\mathbf{0}$ is the closure under (RTE) of the set $\{\langle \Gamma, \Theta \rangle : \Gamma, \Theta \subseteq \text{Form} \,\&\, \Gamma \vDash \Theta\}$. It is easy to see – cf. the proof of Lemma 5.4.1 – that if $\Gamma \vDash_{\text{RTE}} \Theta$, then also $\Gamma \vdash_{\mathbf{0}} \Theta$. On the other hand, $\mathbf{0}$ is the smallest classical logic, so also the converse inclusion obtains. ∎

After these preparations we are ready to proceed to the analysis of particular languages of modal logic. As explained in Section 6.2, if distinctions between different sets of Boolean operators are disregarded, we may say that $\mathcal{L}_\phi$ was the first modal language in actual use, while later $\mathcal{L}_\diamond$ and $\mathcal{L}_\square$ have dominated the scene. The language $\mathcal{L}_\oplus$ seems never to have been used, but this is a historical accident: in every theoretical aspect, it is the equal of the others. The *intended translations* between these four languages are obvious but bear repeating.

$$f_1(\square A) = \neg \Diamond \neg f_1(A), \quad \text{for all } A \in \text{Form}_\square,$$
$$g_1(\Diamond B) = \neg \square \neg g_1(B), \quad \text{for all } B \in \text{Form}_\diamond.$$

$$f_2(\square A) = \neg \phi f_2(A), \quad \text{for all } A \in \text{Form}_\square,$$
$$g_2(\phi B) = \neg \square g_2(B), \quad \text{for all } B \in \text{Form}_\phi,$$

$$f_3(\square A) = \Diamond \neg f_3(A), \quad \text{for all } A \in \text{Form}_\square,$$
$$g_3(\Diamond B) = \square \neg g_3(B), \quad \text{for all } B \in \text{Form}_\diamond.$$

$$f_4(\Diamond A) = \phi \neg f_4(A), \quad \text{for all } A \in \text{Form}_\diamond,$$
$$g_4(\phi B) = \Diamond \neg g_4(B), \quad \text{for all } B \in \text{Form}_\phi.$$

$$f_5(\Diamond A) = \neg \phi f_5(A), \quad \text{for all } A \in \text{Form}_\diamond,$$
$$g_5(\phi B) = \neg \Diamond g_5(B), \quad \text{for all } B \in \text{Form}_\phi.$$

$$f_6(\phi A) = \neg \Diamond \neg f_6(A), \quad \text{for all } A \in \text{Form}_\phi,$$
$$g_6(\Diamond B) = \neg \phi \neg g_6(B), \quad \text{for all } B \in \text{Form}_\diamond.$$

Here we assume that each map h *respects propositional letters* in the sense that it agrees with the identity function on the set of propositional letters, and also that it *respects Boolean operators* in the sense that $h(\oplus[A_0, \ldots, A_{n-1}]) = \oplus[hA_0, \ldots, hA_{n-1}]$, for every Boolean operator $\oplus$ in the language. It is obvious what the intended domain and range of each function is –
$f_1 : \text{Form}_\square \longrightarrow \text{Form}_\diamond$, $g_1 : \text{Form}_\diamond \longrightarrow \text{Form}_\square$, etc. Note that in classical logics these translations agree with those indicated in Section 6.2.

THEOREM 6.3.3. *The lattices of classical logics in the languages $\mathcal{L}_\Box$, $\mathcal{L}_\Diamond$, $\mathcal{L}_\oplus$, and $\mathcal{L}_\phi$ are isomorphic under the maps induced by the intended translations.*

Proof. In view of Theorem 6.3.1 it is enough to prove, for instance, that $\mathbf{0}_\Box$ and $\mathbf{0}_\Diamond$ are syntactically equivalent with respect to f_1 and g_1, $\mathbf{0}_\Diamond$ and $\mathbf{0}_\oplus$ syntactically equivalent with respect to f_4 and g_4, and $\mathbf{0}_\oplus$ and $\mathbf{0}_\phi$ syntactically equivalent with respect to f_6 and g_6. We give a proof of the first claim as an example; the other two are proved similarly.

We begin with some preliminary observations. As a first step, note that

(1) $\qquad\qquad A \sim^* g_1 f_1 A, \qquad$ for all $A \in \text{Form}_\Box$,

(2) $\qquad\qquad B \sim^* f_1 g_1 B, \qquad$ for all $B \in \text{Form}_\Diamond$,

where again $\sim^*$ is the relation of replacement of tautologous equivalents introduced in Section 5.1. We leave the inductive proof as an exercise.

Next we introduce the following piece of terminology. Let $v_\Box$ and $v_\Diamond$ be (RTE)-respecting truth-value assignments for the sets of Boolean atoms in $\mathcal{L}_\Box$ and $\mathcal{L}_\Diamond$, respectively. Then we shall say that $v_\Box$ and $v_\Diamond$ are *coupled* if

(a) $\qquad v_\Box(P) = v_\Diamond(P), \qquad$ for all propositional letters P,

(b) $\qquad v_\Box(\Box A) \neq v_\Diamond(\Diamond \neg f_1 A), \qquad$ for all $A \in \text{Form}_\Box$,

(c) $\qquad v_\Box(\Box \neg g_1 B) \neq v_\Diamond(\Diamond B), \qquad$ for all $B \in \text{Form}_\Diamond$.

Note that conditions (b) and (c) are actually equivalent. To see this, suppose that (b) is true. By (b), $v_\Box(\Box \neg g_1 B) \neq v_\Diamond(\Diamond \neg f_1 \neg g_1 B)$. From the definition of f_1 it follows that $\Diamond \neg f_1 \neg g_1 B = \Diamond \neg \neg f_1 g_1 B$. Using (2) above we conclude that $\Diamond \neg \neg f_1 g_1 B \sim^* \Diamond B$. Therefore, since $v_\Diamond$ is (RTE)-respecting, $v_\Diamond(\Diamond \neg f_1 \neg g_1 B) = v_\Diamond(\Diamond B)$. Consequently, (c) is true. This argument shows that (b) implies (c). That (c) implies (b) is shown in a similar manner.

The interest of the preceding definition should be obvious: if $v_\Box$ and $v_\Diamond$ are coupled (RTE)-respecting truth-value assignments in $\mathcal{L}_\Box$ and $\mathcal{L}_\Diamond$, respectively, then

(3) $\qquad\qquad \bar{v}_\Box(A) = \bar{v}_\Diamond(f_1 A), \qquad$ for all $A \in \text{Form}_\Box$,

(4) $\qquad\qquad \bar{v}_\Box(g_1 B) = \bar{v}_\Diamond(B), \qquad$ for all $B \in \text{Form}_\Diamond$.

Finally, note that whenever $v_\Box$ is an (RTE)-respecting truth-value assignment for the set of Boolean atoms in $\mathcal{L}_\Box$, then there exists an (RTE)-respecting truth-value assignment for the set of Boolean atoms in $\mathcal{L}_\Diamond$ such that $v_\Box$ and $v_\Diamond$ are coupled; and *vice versa*.

With this groundwork we are in a position to prove the theorem. By a result on p. 44, in order to prove that $\mathbf{0}_\Box$ and $\mathbf{0}_\Diamond$ are syntactically equivalent with respect to f_1 and g_1 it would be enough to establish that, for all $\Gamma, \Theta \subseteq \text{Form}_\Box$ and $\Delta, \Lambda \subseteq \text{Form}_\Diamond$,

$$\Gamma \vdash_{\mathbf{0}_\square} g_1 \Lambda \quad \text{iff} \quad f_1\Gamma \vdash_{\mathbf{0}_\diamond} \Lambda,$$

$$g_1\Delta \vdash_{\mathbf{0}_\square} \Theta \quad \text{iff} \quad \Delta \vdash_{\mathbf{0}_\diamond} f_1\Theta.$$

But these conditions are implied by (3), (4), and Lemma 6.3.2. ∎

The preceding results are encouraging. For example, let $S3_\square$ and $S3_\diamond$ be the logics in $\mathcal{L}_\square$ and $\mathcal{L}_\diamond$ that one would intuitively think of as S3. Not only are they syntactically equivalent with respect to f_1 and g_1, but in the sense of the theorem their position within the lattice of classical logics in each language is the same. Moreover, if $S3_{\square\diamond}$ is the logic in the combined language $\mathcal{L}_{\square\diamond}$ that one would intuitively think of as S3, then it can be shown that $S3_{\square\diamond}$ is a conservative definitional extension of both $S3_\square$ and $S3_\diamond$. In particular, in $S3_{\square\diamond}$, for all $A \in \text{Form}_{\square\diamond}$,

$$\square A \dashv\vdash \neg\diamond\neg A, \qquad \diamond A \dashv\vdash \neg\square\neg A.$$

All this seems to fit informal expectations quite well.

Let us see, then, if this analysis can be extended to modal languages based on operators of higher rank than 1. The simplest such operators are of course the binary ones; and the most popular among those have been ⥽, strict implication, and ○, consistency. Let us therefore compare $\mathcal{L}_⥽$ and $\mathcal{L}_\circ$ with the previous modal languages. The intended translations between $\mathcal{L}_\square$ and $\mathcal{L}_⥽$ — as before, respecting propositional letters and Boolean operators — are given by the conditions

$$f_7(\square A) = \top ⥽ f_7(A), \qquad \text{for all } A \in \text{Form}_\square,$$

$$g_7(B ⥽ C) = \square(g_7(B) \to g_7(C)), \qquad \text{for all } B, C \in \text{Form}_⥽;$$

those between $\mathcal{L}_\diamond$ and $\mathcal{L}_\circ$ by

$$f_8(\diamond A) = f_8(A) \circ f_8(A), \qquad \text{for all } A \in \text{Form}_\diamond,$$

$$g_8(B \circ C) = \diamond(g_8(B) \wedge g_8(C)), \qquad \text{for all } B, C \in \text{Form}_⥽.$$

Along the lines of Theorem 6.3.3 we might now try to prove that the lattices of classical logics in $\mathcal{L}_\square$ and $\mathcal{L}_⥽$ are isomorphic under the map Φ_7 induced by f_7 and g_7, and, similarly, that the lattices of classical logics in $\mathcal{L}_\diamond$ and $\mathcal{L}_\circ$ are isomorphic under the map Φ_8 induced by f_8 and g_8. However, interestingly enough, both attempts would fail, and for similar reasons.

In the former case the argument goes like this. Let P and Q be propositional letters. Then $\square(P \to Q) \dashv\vdash \square(\top \to (P \to Q))$ holds in $\mathbf{0}_\square$, and so, by the definition of Φ_7, $P ⥽ Q \dashv\vdash \top ⥽ (P \to Q)$ must hold in $\Phi_7(\mathbf{0}_\square)$. However, by Lemma 6.3.2 $P ⥽ Q \dashv\vdash \top ⥽ (P \to Q)$ does not hold in $\mathbf{0}_⥽$, and in fact fails both ways — it is easy to find (RTE)-respecting truth-value assignments that make one formula true and the other false. Consequently, $\Phi_7(\mathbf{0}_\square) \neq \mathbf{0}_⥽$. In other words, the image under Φ_7 of the lattice of classical logics in $\mathcal{L}_\square$ is not the full lattice of classical logics in $\mathcal{L}_⥽$ (even though of course it is a lattice of classical logics in $\mathcal{L}_⥽$).

A similar analysis applies to $\mathcal{L}_\Diamond$ and $\mathcal{L}_\circ$. Here it turns out that the condition $P \circ Q \dashv \vdash (P \wedge Q) \circ (P \wedge Q)$ holds in $\Phi_8(\mathbf{0}_\Diamond)$ but fails both ways in $\mathbf{0}_\circ$. Thus again the image under Φ_8 of the lattice of classical logics in $\mathcal{L}_\Diamond$ is not the full lattice of classical logics in $\mathcal{L}_\circ$.

Obviously the situation calls for an analysis in general terms. In the following section we sketch the beginnings of such an analysis.

6.4. Two theorems on translations

Up till now we have employed certain typographical symbols that are familiar from modal logic. This choice of symbolism would be a liability in further discussion: the well-known symbols stand in the way of a review of the situation in general terms. It should be noted that so far we have not presupposed anything about the modal operators but only about how they are interrelated, and so it may well be asked in what sense it has been modal logic that we have been dealing with. Thus Theorem 6.3.3 may be — and perhaps should be — seen as a result concerning truth-value functionally complete languages with a single unary non-Boolean operator which are related in certain ways. Or perhaps even better: concerning one such language and certain translations into itself. Similarly, on a deeper level our failure to extend this result to languages involving a binary non-Boolean operator concerns the relationship between two truth-value functionally complete languages, one having a single unary non-Boolean operator, the other having a single binary non-Boolean operator.

These matters can be described in even more general terms. Let the (*non-Boolean*) *profile* of a propositional language be the sequence $\mathbf{z} = \langle z_i \rangle_{i < \omega}$, where, for each $i < \omega$, z_i is the number of primitive i-ary non-Boolean operators of the language, either a natural number or ω. Since we allow ω as a value of z_i, this concept of profile generalizes that of a Boolean profile introduced in Section 3.7. This definition means of course that only denumerable languages have profiles in this sense.

If we ignore the shapes of propositional operators, we might meaningfully introduce the concept of *the language of profile* $\mathbf{z}$, where $\mathbf{z}$ is any non-Boolean profile; this language, then, is uniquely defined up to the choice of propositional letters and Boolean operators, matters on which conventions were adopted in Section 6.3. Consequently, it also becomes meaningful to introduce the concept of *the lattice of classical logics in the language of profile* $\mathbf{z}$. Thus the discussion in Section 6.3 should really be seen as a discussion of the relationship of the lattice of classical logics in the language of profile $\langle 1, 0, 0, 0, \ldots \rangle$ to itself (the existence of several non-trivial automorphisms of a certain kind was established) and to the lattice of classical logics in the language of profile $\langle 0, 1, 0, 0, 0, \ldots \rangle$ (a lack of isomorphisms of a certain kind was indicated but not proved).

Automorphisms and isomorphisms 'of a certain kind' — they would have to be maps induced by translations between the languages, the translations again being of a certain kind. In this book we have often considered translations between languages, and in our general discussion we have been almost excessively liberal: no conditions have been imposed on such functions. However, the particular translations considered in Section 6.3 are strikingly simple. Among other things they all strongly respect substitution, in the following sense. Let $f : \text{Form}_1 \longrightarrow \text{Form}_2$ be a map from the set of formulas of a language $\mathcal{L}_1$ to the set of formulas of a language $\mathcal{L}_2$. Suppose that whenever $\oplus$ is an n-ary operator of $\mathcal{L}_1$ then there is some formula $D_\oplus \in \text{Form}_2$ containing at least n propositional letters $P_0, \ldots, P_{n-1}$ such that, for all formulas $A_0, \ldots, A_{n-1} \in \text{Form}_1$,

$$f(\oplus[A_0, \ldots, A_{n-1}]) = D_\oplus(P_0/fA_0, \ldots, P_{n-1}/fA_{n-1}).$$

Then we say that f *respects substitution*. If the preceding condition holds with 'at least' replaced by 'exactly', then we say that f *strongly respects substitution*. Notice that the functions $\longrightarrow$ and $\longleftarrow$ defined in Section 5.5 respect propositional letters and substitution — substitution not necessarily strongly, though.

We can now prove a theorem which seems to extract the general idea implicit in the two instances concerning $\mathcal{L}_\square$ and $\mathcal{L}_{\rightarrow}$ on the one hand and $\mathcal{L}_\lozenge$ and $\mathcal{L}_\circ$ on the other.

THEOREM 6.4.1. *Let $\mathcal{L}_1$ and $\mathcal{L}_2$ be the languages of profile $\langle n, 0, 0, 0, \ldots \rangle$ and $\langle 0, 1, 0, 0, 0, \ldots \rangle$, respectively, where n is either a positive natural number or else ω. Let $\mathbf{L}_1$ and $\mathbf{L}_2$ be the lattices of classical logics in each. Then there are no translations $f : \text{Form}_1 \longrightarrow \text{Form}_2$ and $g : \text{Form}_2 \longrightarrow \text{Form}_1$ respecting propositional letters and substitution such that $\mathbf{L}_1$ and $\mathbf{L}_2$ are isomorphic under the map induced by f and g.*

Proof. First let us introduce the following new terminology. We say that a formula A of any language *reduces to* a formula B of the same language if $A \dashv\vdash_{\text{RTE}} B$ and $l(A) \geq l(B)$ (for the length of an expression, see p. 24), and that A is *reduced* if it reduces to itself. Evidently every formula reduces to some formula or other. Moreover, this concept of reduction is a transitive relation.

Suppose now that $f : \text{Form}_1 \longrightarrow \text{Form}_2$ and $g : \text{Form}_2 \longrightarrow \text{Form}_1$ are translations which respect propositional letters and strongly respect substitution. This is a stronger assumption than warranted by the formulation of the theorem, but it simplifies the proof somewhat; we leave the more general case as an exercise. Let ☆ be the sole non-Boolean operator of $\mathcal{L}_2$. Since g strongly respects substitution there is some reduced formula $D \in \text{Form}_1$ containing exactly two propositional letters P_0 and P_1 such that

$$g(B ☆ C) = D(P_0/gB, P_1/gC), \quad \text{for all } B, C \in \text{Form}_2.$$

Let $\oplus_0, \ldots, \oplus_{k-1}$ be all the non-Boolean operators occurring in D. Since also f strongly respects substitution there is, for each $i < k$, a reduced formula

$C_i \in \text{Form}_2$ containing exactly one propositional letter P_{i+2} such that

$$f(\oplus_i A) = C_i(P_{i+2}/fA), \text{ for all } A \in \text{Form}_1.$$

Note that, since f respects propositional letters, $f(\oplus_i P_{i+2}) = C_i$.

We should like to prove that $\mathbf{L_1}$ and $\mathbf{L_2}$ are not isomorphic under the map induced by f and g. Preparing for *contradictio*, suppose that they are. Then $\mathbf{0_1}$ and $\mathbf{0_2}$ must be syntactically equivalent with respect to f and g. Let P and Q be any distinct propositional letters. By Lemma 6.3.2, then, $P \star Q \dashv\vdash_{RTE}$ $fg(P \star Q)$. Consequently, there must be some $Y, Z \in \text{Form}_2$ such that $Y \star Z$ occurs in $fg(P \star Q)$ and

(1) $\qquad\qquad\qquad$ Y reduces to P

(2) $\qquad\qquad\qquad$ Z reduces to Q.

There must also be some $X \in \text{Form}_1$ and some $i < k$ such that, for some expressions E, F of $\mathcal{L}_1$, $f(\oplus_i X) = E \star (Y \star Z) \star F$. It follows that there are formulas $Y^\dagger, Z^\dagger$ and expressions $E^\dagger, F^\dagger$ of $\mathcal{L}_2$ such that

(3) $\qquad\qquad\qquad f(\oplus_i P_{i+2}) = E^\dagger \star (Y^\dagger \star Z^\dagger) \star F^\dagger,$

(4) $\qquad\qquad\qquad Y = Y^\dagger(P_{i+2}/fX),$

(5) $\qquad\qquad\qquad Z = Z^\dagger(P_{i+2}/fX).$

Consider $Y^\dagger$. By (3), $Y^\dagger$ occurs in C_i, and so, since f strongly respects substitution, $Y^\dagger$ contains no propositional letter other than P_{i+2}. (This is the only step in the proof which really depends on the assumption that f and g strongly respect substitution and which a general argument, not assuming strength, will have to modify.) Therefore $Y^\dagger$ must reduce to one of $P_{i+2}, \neg P_{i+2}, \top,$ and $\bot$ (if $\top$ or $\bot$ is not primitive, we assume that it stands for some tautology, respectively, contradiction that is reduced). Using (1) and (4) we see that if $Y^\dagger$ reduces to P_{i+2}, then $fX \dashv\vdash_{RTE} P$; that if $Y^\dagger$ reduces to $\neg P_{i+2}$, then $fX \dashv\vdash_{RTE} \neg P$; but that it is impossible that $Y^\dagger$ should reduce to either $\top$ or $\bot$. We conclude that

(6) $\qquad\qquad$ either $fX \dashv\vdash_{RTE} P$ or $fX \dashv\vdash_{RTE} \neg P$.

Considering $Z^\dagger$ instead, we arrive by a similar route via (2) and (5) at the conclusion

(7) $\qquad\qquad$ either $fX \dashv\vdash_{RTE} Q$ or $fX \dashv\vdash_{RTE} \neg Q$.

By (6) and (7), then, $P \dashv\vdash_{RTE} Q$ or $P \dashv\vdash_{RTE} \neg Q$. As P and Q are distinct, this is an absurd result. ∎

The study of the relationship between the lattices of classical logics in languages of different profiles and under varying conditions on the translations involved may well be interesting. Here is another sample result, generalizing a theorem in Chellas (1975), which it is instructive to compare with the preceding theorem:

THEOREM 6.4.2. *Assuming denumerable languages, the lattices of classical logics in the languages of profiles $\langle \omega, 0, 0, 0, \ldots \rangle$ and $\langle 0, 1, 0, 0, 0, \ldots \rangle$ are isomorphic under a map induced by translations respecting propositional letters and Boolean operators.*

Proof. Let $\mathcal{L}_1$ be the language of profile $\langle \omega, 0, 0, 0, \ldots \rangle$ and $\mathcal{L}_2$ that of the profile $\langle 0, 1, 0, 0, 0, \ldots \rangle$. Suppose that $\oplus_0, \oplus_1, \ldots, \oplus_i, \ldots$ is an exhaustive enumeration without repetitions of the non-Boolean operators in $\mathcal{L}_1$, and suppose that $A_0, A_1, \ldots, A_i, \ldots$ is an enumeration of the formulas of $\mathcal{L}_1$, also exhaustive and non-repetitive. (It is here that the assumption of denumerable languages comes in. We have already assumed that the set of propositional letters is denumerable, and the set of non-Boolean operators is denumerable by hypothesis; so in effect what we have added is the assumption that also the set of Boolean operators is at most denumerable — a modest assumption.) We associate with each formula $A \in \text{Form}_1$ an operator $\oplus_A$ of $\mathcal{L}_1$ in the following way. For any n, suppose that $\oplus_{A_0}, \ldots, \oplus_{A_{n-1}}$ have already been defined, and that they are the same operators as $\oplus_0, \ldots, \oplus_{p-1}$, for some p. Then $\oplus_{A_n}$ is defined as follows. If $A_n \sim^* A_k$, for some $k < n$ (here $\sim^*$ is the relation of replacement of tautological equivalents defined in Section 5.1), then $\oplus_{A_n} = \oplus_{A_k}$. On the other hand, if there is no $k < n$ such that $A_n \sim^* A_k$, then $\oplus_{A_n} = \oplus_p$. Evidently it follows from this definition that, for all $A, B \in \text{Form}_1$, $\oplus_A = \oplus_B$ if and only if $A \sim^* B$.

Suppose that $\star$ is the sole operator of $\mathcal{L}_2$, hence binary. We define translations $f : \text{Form}_1 \longrightarrow \text{Form}_2$ and $g : \text{Form}_2 \longrightarrow \text{Form}_1$, respecting propositional letters and Boolean operators, as follows:

$$f(\oplus_A B) = fA \star fB, \quad \text{for all } A, B \in \text{Form}_1,$$

$$g(A \star B) = \oplus_{gA} gB, \quad \text{for all } A, B \in \text{Form}_2.$$

Note that f and g are particularly nicely related:

$$gfA = A, \quad \text{for all } A \in \text{Form}_1,$$

$$fgB = B, \quad \text{for all } B \in \text{Form}_2.$$

By Theorem 6.3.1, in order to show that $\mathbf{L}_1$ and $\mathbf{L}_2$ are isomorphic under f and g, it is now enough to show that $\mathbf{0}_1$ and $\mathbf{0}_2$ are syntactically equivalent with respect to f and g. To do this we proceed as in the proof of Theorem 6.3.3. Let us say that two (RTE)-respecting truth-value assignments in $\mathcal{L}_1$ and $\mathcal{L}_2$, respectively, are *coupled* if they agree on the set of propositional letters and

$$v_1(\oplus_A B) = v_2(fA \star fB), \quad \text{for all } A, B \in \text{Form}_1,$$

$$v_1(\oplus_{gA} gB) = v_2(A \star B), \quad \text{for all } A, B \in \text{Form}_2.$$

(Actually, the last two conditions are equivalent: one holds if and only if the other does.) The rest of the argument is analogous to that in the proof of Theorem 6.3.3. ∎

Evidently the translations f and g of the last proof satisfy much stronger conditions than those mentioned in the statement of the theorem, respect of propositional letters and Boolean operators: f strongly respects substitution, and even though g does not respect substitution, it exhibits a kind of partial respect. It is clear that the infinite supply of primitive operators in $\mathcal{L}_1$ is crucial to the proof, and so there should be a way to formulate the theorem which would make it false if ω were replaced by any finite n. At any rate, as g does not respect substitution, any such reformulation would be consistent with Theorem 6.4.1.

6.5. A definition of modal logic

In previous chapters we were able to characterize important classes of logics in very general terms. A virtue of those definitions was their neutrality between individual object languages. It would be gratifying if an adequate similar definition of an important class of modal logics could be found; such was originally the author's hope (cf. the Preface and Section 6.1). But such a definition, which of course would not mention particular operators like □ or ◊, has proved difficult to find; at least our analysis has not resulted in one. As we saw in the preceding sections, to define modal logic as, for example, classical logic with one non-Boolean operator will not do; that definition would be too wide, but, more seriously, the lattices of such logics vary with the profile of the language. The latter difficulty could be removed by defining modal logic as classical logic of one unary non-Boolean operator — this concept could be made meaningful even over languages with other kinds of non-Boolean operators — but the objection that such a definition is too wide would remain. Perhaps semantics holds more promise for a search for a general definition. However, semantic methods for dealing with very weak, non-congruential logics would first have to be developed. Besides, the characterization problem would recur in one form or other: in the final analysis, when is a frame a modal frame, an algebra a modal algebra?

Thus one may have to accept the idea that reference to particular operators must be made. For historical reasons it would be most appropriate to single out either □ or ◊ for such reference. This is what we will now do. Attempting to seize on a structural condition that captures what is basic to the slightly fuzzy set of systems which go under the name of modal logic, in the naïve sense of the word, we propose that a logic in $\mathcal{L}_\square$ be called *modal* if it satisfies the condition

if $\Gamma \vDash A,$

then $\square\Gamma \vdash \square A,$ *provided that* $\Gamma \neq \emptyset$;

or, dually, that a logic in $\mathcal{L}_\diamond$ be called *modal* if it satisfies the condition *if* B ⊨ Θ,

then ◊B ⊢ ◊Θ, *provided that* Θ ≠ ∅.

Furthermore, we say that any logic L in any language $\mathcal{L}$ is *modal (under f and g)* if f and g are translations respecting propositional letters and substitution and either (i) f : Form ⟶ Form$_\square$ and g : Form$_\square$ ⟶ Form and there is a modal logic L$_\square$ in $\mathcal{L}_\square$ with which L is syntactically equivalent with respect to f and g, or (ii) f : Form ⟶ Form$_\diamond$ and g : Form$_\diamond$ ⟶ Form and there is a modal logic L$_\diamond$ in $\mathcal{L}_\diamond$ with which L is syntactically equivalent with respect to f and g.

Similarly, let us say that a logic is normal in the sense of Tarski, or *Tarski normal (Tarski normal under f and g)*, if it satisfies the appropriate conditions with the provisos specifying non-emptiness replaced by provisos specifying emptiness.

We observe that the following familiar entailments hold in any classical modal logic:

$\square(A \to B), \square A \vdash \square B$ $\diamond A \vdash \diamond B, \diamond(A \land \neg B)$

$\square(A \land B) \vdash \square A$ $\diamond A \vdash \diamond(A \lor B)$

$\square(A \land B) \vdash \square B$ $\diamond B \vdash \diamond(A \lor B)$

$\square A, \square B \vdash \square(A \land B)$ $\diamond(A \lor B) \vdash \diamond A, \diamond B$

Similarly, in any Tarski normal classical logic,

$\vdash \square \top$ $\diamond \bot \vdash$.

The reader will recall that in traditional modal logic one defines a logic as *regular* if the so-called Scott's Rule is satisfied:

if Γ ⊢ A,

then □Γ ⊢ □A, *provided that* Γ ≠ ∅;

or, dually,

if B ⊢ Θ,

then ◊B ⊢ ◊Θ, *provided that* Θ ≠ ∅.

Similarly, a logic is defined as normal in the sense of Lemmon, or just *normal*, if these conditions hold even without the provisos of non-emptyness. Thus a normal logic is closed under the rules ⊢A/⊢□A and B⊢/◊B⊢ ('if A is a thesis, then so is □A'; 'if B is an antithesis, then so is ◊B'). These definitions are easily integrated into the present framework. This means that within any given truth-value functionally complete language with at least one non-Boolean operator, and given translations of the right kind, we have, among others, nine classes of logics which are represented in the following diagram:

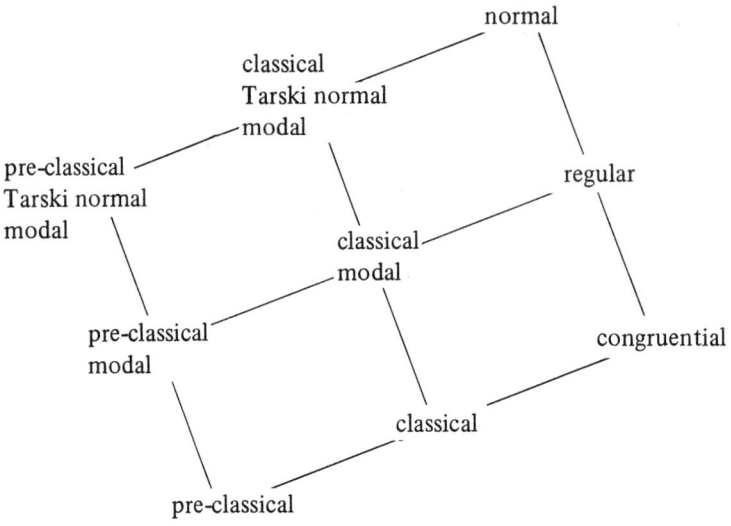

The way to read the diagram is this: if, and only if, a logic falls under one of these headings, it also falls under any heading that can be reached by way of an unbroken downward line. Moreover, a pre-classical logic is regular if and only if modal and congruential; normal if and only if it is Tarski normal, modal and congruential. This classification, which departs from Segerberg (1971), makes S4 and S5 classical normal modal logics, and S2 and S3 classical Tarski normal modal logics. Virtually no system in the literature is just pre-classical modal or pre-classical Tarski normal modal, but the point of Makinson (1973) is that such systems exist; and remember that Makinson's Warning applies to all logics that are not classical, hence also to non-classical modal logics! Very few systems are just classical modal – C0.5 and D0.5 are among the few that are. The famous but deviant S1 becomes classical but not modal; at first this may seem to tell against our proposal, but perhaps it may be argued that 'classical but not modal' is not a bad description of that little understood system.

Let it be emphasized that while 'normal', 'regular' and 'congruential' are part of reasonably standard terminology, the remaining concepts in the diagram are not. As the name indicates, Tarski normality goes back to a related concept (in modal algebra) due to A. Tarski, but even that is not a standard concept in modal logic. The other concepts – pre-classical, classical, modal – are proposed for the first time in this book. Whether our proposal is fruitful remains to be seen; there certainly are other candidates that one may think of. However, some technical definition of modal logic is needed in order for the question at the end of Section 6.2 to be answerable. Our proposal is at least an example of the kind of *ad hoc* structural definition to which we may have to resort in the absence of some more satisfying analysis.

Thus we are left with a programme instead of an analysis. Playing on the double meaning of words, informal versus formal, what it amounts to is that modal logic is the study of modal logics, and that classical modal logic is the study of classical modal logics. Apart from its arbitrariness, this is a point which would have realized the original monograph ambition described in Section 6.1: it is from such a point that an essay on classical modal logic should take off.

REFERENCES

Aczel, P. H. G. (1967). Some results on intuitionistic predicate logic. *J. symbolic Logic*, **32**, 556.
Anderson, A. R. and Belnap, N. D., Jr. (1975). *Entailment. The logic of relevance and necessity*, Vol. I. Princeton University Press, Princeton, New Jersey.
Åqvist, L. (1973). Modal logics with subjunctive conditionals and dispositional predicates. *J. phil. Logic*, **2**, 1–76.
Boolos, G. (1979). *The unprovability of consistency: an essay in modal logic.* Cambridge University Press, Cambridge.
Chellas, B. F. (1975). Basic conditional logic. *J. phil. Logic*, **4**, 133–53.
—— (1980). *Modal logic: an introduction.* Cambridge University Press, Cambridge.
Cresswell, M. J. (1967). A Henkin completeness theorem for T. *Notre Dame J. formal Logic*, **8**, 186–90.
Enderton, H. B. (1972). *A mathematical introduction to logic.* Academic Press, New York.
Fine, K. (1974). Models for entailment. *J. phil. Logic*, **3**, 347–72.
Gödel, K. (1933). Eine Interpretation des intuitionistischen Aussagenkalküls. *Ergebnisse eines mathematischen Kolloquiums*, **4**, 39–40.
Goldblatt, R. (1974). Semantic analysis of orthologic. *J. phil. Logic*, **3**, 19–35.
Halldén, S. (1949). A reduction of the primitive symbols of the Lewis calculi. *Portugaliae mathematica*, **8**, 85–8.
Halmos, P. R. (1960). *Naive set theory.* Van Nostrand, Princeton, New Jersey.
Henkin, L. (1949). The completeness of the first-order functional calculus. *J. symbolic Logic*, **14**, 159–66.
Heyting, A. (1956). *Intuitionism: an introduction.* North-Holland, Amsterdam.
Hughes, G. E. and Cresswell, M. J. (1968). *An introduction to modal logic.* Methuen, London.
Kanger, S. (1957). *Provability in logic.* Stockholm Studies in Philosophy, no. 1. Almqvist & Wiksell, Stockholm.
—— (1968). Equivalent theories. *Theoria*, **34**, 1–6.
Kelley, J. L. (1955). *General topology.* Van Nostrand, New York.
Kleene, S. C. (1952). *Introduction to metamathematics.* North-Holland, Amsterdam.
Kneebone, G. T. (1963). *Mathematical logic and the foundations of mathematics.* Van Nostrand, London.
Kotas, J. and Pietczkowski, A. (1970). Allgemeine logische und mathematische Theorien. *Z. math. Logik Grundlagen Math.*, **16**, 353–76.
Kripke, S. A. (1963). Semantical analysis of modal logic I. Normal modal propositional calculi. *Z. math. Logik Grundlagen Math.*, **9**, 67–96.
—— (1965). Semantical analysis of modal logic II. Non-normal modal propositional calculi. In *The theory of models* (ed. J. W. Addison, L. Henkin, and A. Tarski), pp. 206–20. North-Holland, Amsterdam.

REFERENCES

Kuhn, S. T. (1976). *Many-sorted modal logics*. Philosophical Studies published by the Philosophical Society and the Department of Philosophy, University of Uppsala, no. 29.

Lemmon, E. J. (1957). New foundations for Lewis modal systems. *J. symbolic Logic*, 22, 176-86.

— (1977). *The 'Lemmon notes': an introduction to modal logic*. (In collaboration with Dana Scott.) *American Philosophical Quarterly* monograph series, no. 11. Blackwell, Oxford.

Lewis, C. I. (1912). Implication and the algebra of logic. *Mind*, n.s., 21, 522-31.

— (1918). *A survey of symbolic logic*. University of California Press, Berkeley.

Lewis, C. I. and Langford, C. H. (1932). *Symbolic logic*. Second edition 1959. Dover Publications, New York.

Makinson, D. (1966). On some completeness theorems in modal logic. *Z. math. Logik Grundlagen Math.*, 12, 379-84.

— (1973). A warning about the choice of primitive operators in modal logic. *J. phil. Logic*, 2, 193-6.

Montgomery, H. and Routley, R. (1966). Contingency and non-contingency bases for normal modal logics. *Logique et analyse*, n.s., 9, 318-28.

Prawitz, D. (1965). *Natural deduction: a proof-theoretical study*. Stockholm Studies in Philosophy, no. 3. Almqvist & Wiksell, Stockholm.

Rasiowa, H. and Sikorski, R. (1963). *The mathematics of metamathematics*. Panstwowe Wydawnictwo Naukowe, Warsaw.

Routley, R. and Meyer, R. K. (1972a). The semantics of entailment. I. In *Truth, syntax and modality* (ed. H. Leblanc), pp. 199-243. North-Holland, Amsterdam.

— and — (1972b). The semantics of entailment. II. *J. phil. Logic*, 1, 53-73.

— and — (1972c). The semantics of entailment. III. *J. phil. Logic*, 1, 192-208.

Scott, D. (1971). On engendering an illusion of understanding. *J. Phil.*, 68, 787-807.

— (1972). Background to formalization. I. In *Truth, syntax and modality* (ed. H. Leblanc), pp. 244-73. North-Holland, Amsterdam.

— (1974). Completeness and axiomatizability in many-valued logic. In *Proceedings of the Tarski Symposium* (ed. L. Henkin et al.). Proceedings of symposia in pure mathematics. The American Mathematical Society, Providence, Rhode Island.

Segerberg, K. (1968). Propositional logics related to Heyting's and Johansson's. *Theoria*, 34, 26-61.

— (1971). *An essay in classical modal logic*. Philosophical Studies published by the Philosophical Society and the Department of Philosophy, University of Uppsala, no. 13.

Takeuti, G. (1975). *Proof theory*. North-Holland, Amsterdam.

Thomason, R. H. (1968). On the strong semantical completeness of the intuitionistic predicate logic. *J. symbolic Logic*, 33, 1-7.

von Wright, G. H. (1951). *An essay in modal logic*. North-Holland, Amsterdam.

INDEX OF TERMS

abbreviatory device 94, 95
agree 32
alphabet 23
ambition 125
ancestral 11
antecedent 67
antisymmetric 16
antithesis 35, 139
arity 6
assertivity 39
assignment of truth-values 53
associated condition 68
atom, Boolean 51
available 94
avoids 36, 49
Axiom of Choice 49, 61, 65, 108
axiom system 35, 83

backward induction 12
base 128
basic
    object 6
    step 2
beweisbar 127
Big Seven 66, 94
binary 6, 22
Boolean
    atom 51
    combination 51
    compound 51
    degree 77, 79
    extension 89
    logic 62
    matrix 53
    profile 77
    propositional operator 22
    purely 51
box 127
branch 16

calculus of
    consistencies 127
    ordinary inference 127
    strict implication 126

chain 48
    of virtually finite sequences 20
character, finite 47
choice
    axiom of 49, 61, 65, 108
    function 49
classical
    logic (formal sense) 105, 113
    logic (informal sense) 18, 19, 49
    operator 110
closed under
    a relation 7
    cut 38
    replacement of definitional equivalents 45
    replacement of tautological equivalents 109
    substitution 38
closure, reflexive transitive 11
combination, Boolean 51
common logic 38
compact 40
completeness 56, 59, 62, 63, 65, 83
    Post 129
    truth-value functional 92, 130
complex Boolean atom 51
compound, Boolean 51
concatenation 23
condition
    antecedent 67
    associated 68
    consequent 67
    elimination 67, 68, 96
    introduction 67, 69, 97
condition of
    assertivity 39
    consistency 39
    cut 37
    diagonality 36
    finitariness 40
    monotonicity 37
    overlap 36
    reflexivity 36
    substitutivity 38

condition of (*cont.*)
    transitivity 38
conditions, Scott's 97
congruential 41, 63
conjunction 66, 94, 99
    intensional 126
connected 16
consequent 67
conservative 45
consistencies 127
consistency 39
consistent 35, 39, 126
constant
    falsity 66
    propositional 22
    truth 66
constructive 18, 49, 65
contingent 128
contradiction 53, 56
conventions 12, 23, 26, 35, 94, 99
convex 23
coupled 132, 137
course-of-values induction 4
cut 37
    elimination 83
    logic 38

decision procedure 76
deducibility relation 36
deducible 34, 126
deduction theorem 98
defining formula 89
definition 89
definitional 45, 91
degree 77, 79
derived rule 83
device, abbreviatory 94, 95
diagonal 10
diagonality 36
diamond 127
dilution 37
directly express 55
directly expressible 55
disjunction 66, 94, 99
    intensional 126
disproof 83
domain 75
dominance chain 20, 78
dominate 20, 78
downward inductive system 13

either-or 126
elimination condition 67, 68, 96
empty
    relation 6
    sequence 23
    set 6

entailment 34, 36, 128
epistemological 19
equal to, identically 72
equivalence
    material 66, 94
    strict 126
    syntactic 42, 43
equivalent
    definitional 45
    M– 53, 106
    syntactically 43
    tautologically 56
express 85
expressible 85
expression 24
extension
    Boolean 89
    conservative 45
    definitional 45, 91
    of a language 44
    of a logic 45
    of a truth-value assignment 53
external 104

False, the 52
falsity, constant 66, 94
finished 76
finitary 16, 40
finite character 47
finitely 66
finiteness theorem 99
finite, virtually 19
fish-hook 126
formula 25
    defining 89
formulawise 65
function
    choice 49
    substitution 31
    truth-value 53, 66

generated
    object 6
    subchain 20
generating relation 6

Hauptsatz 83

identically equal to 72
iff 10
immediate
    Boolean extension 89
    predecessor 4, 15
    subformula 28
    successor 5, 15
implausible 35
implication

INDEX OF TERMS

M- 53
  material 66, 94, 126
  strict 126
  tautological 56
implicit matrix 56
impossible 126
inconsistent 35, 39
induced
  map 130
  object 9
induction 1, 3
  backward 12
  course-of-values 4
  from the bottom up 9
  from the top down 9
  hypothesis 2
  mathematical 3
  ordinary 3
  over an inductive system 7
  over the set of natural numbers 1, 3
  super-principle of 7
  variable 3
inductive
  logic 37
  step 2
  system 6
inference 127
  rule 83
infinitary 16
initial segment 23
insensitive 100
instance, substitution 32
intended translations 131
intensional
  conjunction 126
  disjunction 126
  propositional operator 22
interdeducible 34
internal 104
introduction condition 67, 69, 97
intuitionism 49
irreflexive 16
isomorphism 115, 116, 134
  theorem 123

König's Lemma 18, 19, 78

language
  meta- 26, 94
  object- 26, 63, 94, 129, 130, 138
  propositional 22
  sensitive 100, 104, 125
  truth-value functionally complete 130
lattice 115
  isomorphism 116
  of classical logics 115
left-assertivity 39

left-monotonicity 37
Lemmon, normal in the sense of 139
length
  of an expression 24
  of a sequence 14
  virtual 20
letter, propositional 22
Lindenbaum's Lemma 47, 50, 60
linear ordering 16
logic 34
  Boolean 62
  classical (formal sense) 105, 113
  classical (informal sense) 18, 49
  common 38
  cut 38
  inconsistent 39
  inductive 37
  in the wide sense 44
  modal (formal sense) 138, 141
  modal (informal sense) vii, 125, 141
  pre-classical 92
  quantum 37
  relevance 37
  truth-value functionally complete 92
  two-valued 52

Makinson's Warning 100, 104, 140
map, induced 130
mark 20
material
  equivalence 66, 94
  implication 66, 94, 126
mathematical induction 3
matrix
  Boolean 53
  implicit 56
maximal
  consistent 50
  for a property 47
M-contradiction 53
M-equivalent 53, 106
meta-language 26, 94
metalevel 4
M-implication 53
modal
  language 130
  logic (formal sense) 138, 141
  logic (informal sense) vii, 125, 141
  operator 127
monograph ambition 125
monoid 24, 32
monotonicity 37
M-tautology 53

natural number 1, 22
necessarily, either-or 126
necessary 127

INDEX OF TERMS

negation 66, 94
node 15
non-Boolean
   operator 22
   profile 134
non-constructive 49
non-contingent 128
non-necessary 128
normal 128, 139
   Tarski 139

object
   basic 6
   generated 6
   induced 9
   language 26, 63, 94, 126, 129, 130, 138
   level 4
occur 23, 28
operator
   Boolean 22
   classical 110
   congruential 41
   intensional 22
   modal 127
   non-Boolean 22
   primitive 100
   propositional 22
or 126
ordering 16
ordinary
   induction 3
   inference 127
overlap 25, 36

parse 26
partial ordering 16
partitioning 57
plausible 35
possible 127
Post complete 129
precede 15
pre-classical 92
predecessor 4, 15
preserve 7
primitive
   operator 100
   rule 83
   symbol 23, 26
principle of
   course-of-values induction 4
   induction over an inductive system 7
   mathematical induction 3
product, relative 10
profile
   Boolean 77
   non-Boolean 134
proof 83

proof-tree 76
property 3, 47, 104
propositional 22
provability 127
purely Boolean 51

quantum logic 37

rank 8
readability 27
reductio ad absurdum 49
reflexivity 16, 36
regular 139
relation 6
   closed under 7
   deducibility 36
   empty 6
   generating 6
relative product 10
relevance logic 37
replacement of
   definitional equivalents 45
   interdeducible formulas 41
   M-equivalents 106
   tautological equivalents 106
respects 56
   Boolean operators 131
   finitely 66
   formulawise 65
   propositional letters 131
   (RTE) 130
   substitution 135
result of substitution 31
right-assertivity 39
right-monotonicity 37
root 16
rule 83
   Scott's 139

Scott's conditions 97
Scott's rule 139
search-tree 75
self-reference 148
semantics 36, 40, 44, 63, 138
semigroup 24
sensitivity 100, 104, 125
sequence 12, 23
   empty 23
   virtually finite 19
Sheffer stroke 128
simultaneous substitution 31
step
   basic 2
   inductive 2
strict implication 126
   calculus of 126
strict linear ordering 16

strict partial ordering 16
subchain 20
subexpression 24
subformula 28
sublanguage 44
sublogic 45
substitution 31
substitutional 38
substitutivity 38
succeed 15
successor 15
super-principle of induction 7
symbol, primitive 23, 26
syntactic equivalence 43
syntax 63
system
   axiom 35, 83
   inductive 6

Tarski normal 139
tautologically
   equivalent 56
   implies 56
tautology 53, 56
ternary 6, 22
textbook ambition 125
theorem 35
   completeness 56, 59, 62, 63, 65
   conservation 92
   deduction 98
   finiteness 99
   isomorphism 123
thesis 35, 100, 139
thinning 37

top 16
transitivity 16, 38
translation 44, 130
   intended 131
tree 12, 75
tree-structure 12
True, the 52
truth, constant 66, 95
truth-table 58
truth-value 52
   function 53, 66
   functionally complete 92, 130
truth-values, assignment of 53
Tukey's Lemma 47, 49, 50
tuple 6
two-valued logic 52
type 57
type-determination 59

unary 22
unique readability 27
upward inductive system 13
use and mention 93

virtual length 20
virtually finite 19

weakening 37
weakly
   connected 16
   interdeducible 34

zeroary 6, 22
Zorn's Lemma 48, 49

INDEX OF SYMBOLS

ω 1
$\emptyset$ 6
$\langle B, R \rangle$ 6
B_* 6
R_{nat} 7
B^* 9
id_A 10
iff 10
■ 10
R^* 11
x, y, z 12, 75
$\langle \alpha_i \rangle_{i<k}$ 17
$\langle \alpha_i \rangle_{i<\omega}$ 17
$\langle \alpha_i \rangle_i$ 17
$\{\alpha_i\}_i$ 17
$\mathcal{L}$ 22
Lett 22
Bop 22
Iop 22
r 22
Op 22
I_n 23
★ 23
Exp 24
l 24
Form 25
$\oplus[A_0,\ldots,A_{n-1}]$ 26
$\oplus$ 26
$\oplus A$ 26
$A \oplus B$ 26
$A(P/B)$ 31
$A(P_0/B_0,\ldots,P_{n-1}/B_{n-1})$ 31
$s\Sigma$ 33

$\langle \Gamma, \Theta \rangle$ 34
$\vdash$ 34
$\dashv\vdash$ 34
$\not\vdash$ 35
Th(L) 35
Antith(L) 35
(Refl) 36
(Diag) 37
(Overl) 37
(Mono) 37
$(Mono_L)$ 37
$(Mono_R)$ 37
(Cut_0) 37
(Cut_1) 37
(Cut_2) 37
(Cut) 37
(Cut_G) 38
(Trans) 38
(Subst) 38
$\dashv$ 38
(Cons) 39
(Ass) 39
(Ass_L) 39
(Ass_R) 39
(Fin) 40
$(\oplus_{df})$ 45
$\widetilde{df}$ 45
$\widetilde{df}^*$ 45
$\{0,1\}^n$ 53
$\mathbf{M}$ 53
$f^{\mathbf{M}}$ 53
$\vdash^{\mathbf{M}}$ 53
$\dashv\vdash^{\mathbf{M}}$ 53

INDEX OF SYMBOLS

$\vDash$ 56
$\dashv\vDash$ 56
$\bar{f}$ 56
$\langle I, J \rangle$ 57
T 66
F 66
N 66
K 66
A 66
C 66
E 66
(**EAT**), etc. 67, 68
(**IAF**), etc. 67, 69
(**EA**$\phi_{\langle I, J \rangle}$) 71
(**EC**$\phi_{\langle I, J \rangle}$) 71
(**IA**ϕ) 71
(**IC**ϕ) 72
$\phi \equiv t$ 72
(**IAF**$_n$) 72
(**ICT**$_n$) 72
(**IAF**$'_n$) 72
(**ICT**$'_n$) 72
$\mathbf{P}°$(Form) 75
dom T 75
$\top$ 94
$\bot$ 94
$\neg$ 94
$\vee$ 94
$\wedge$ 94
$\to$ 94
$\leftrightarrow$ 94
(**EA**$\top$), etc. 96
(**IA**$\bot$), etc. 97
(**S**$\top$), etc. 97, 98

$\widetilde{\mathbf{M}}$ 106
$\widetilde{\mathbf{M}}^*$ 106
$\sim$ 106
$\sim^*$ 106
$\widetilde{\widetilde{\mathbf{M}}}$ 108
$\approx$ 109
(RTE) 109
$\mathbf{1}_{\mathcal{L},\mathbf{M}}$ 114
$\mathbf{1}$ 114
$\mathbf{0}_{\mathcal{L},\mathbf{M}}$ 114
$\mathbf{L}_{\mathcal{L},\mathbf{M}}$ 115
$\longrightarrow$ 116
$\longleftarrow$ 116
ν 126
$\rightarrowtail$ 126
$\lozenge$ 126
$=_{df}$ 126
$=$ 126
$\circ$ 126
$\diamond$ 127
$\square$ 127
$\triangle$ 127
$\triangledown$ 127
$\mathbf{v}$ 128
$\mathbf{\wedge}$ 128
ϕ 128
L 128
M 128
N' 128
A' 128
C' 128
E' 128
$\mathcal{L}_s$ 130
Form$_s$ 130
$\mathbf{0}_s$ 130

INDEX OF NAMES

Aczel, P. H. G. 63, 143
Addison, J. W. 143
Anderson, A. R. 63, 143
Aqvist, L. 38, 143

Belnap, N. D., Jr. 36, 143
Bolzano, B. 19
Boolos, G. 127, 143

Chellas, B. F. 128, 136, 143
Cresswell, M. J. 63, 128, 143

Dugundji, J. 129

Enderton, H. B. 9, 143

Fine, K. 63, 143
Fitch, F. B. 127
Fleming, R. vii

Gentzen, G. 37, 73, 83, 96
Gödel, K. 127, 128, 143
Goldblatt, R. I. 63, 143

Halldén, S. 128, 129, 143
Halmos, P. R. 48, 49, 143
Henkin, L. 63, 65, 143, 144
Heyting, A. 49, 143, 144
Hilpinen, R. viii
Hughes, G. E. 128, 143

Johansson, I. 144

Kalmár, L. 63, 65
Kanger, S. 44, 96, 143
Kaplan, D. 63
Kelley, J. L. 48, 49, 143
Kleene, S. C. 34, 63, 96, 143
Kneebone, G. T. 126, 143
König, J. 18, 19, 78
Kotas, J. 44, 143
Kripke, S. A. 128, 129, 143
Kuhn, S. T. vii, 40, 144

Langford, C. H. 127, 128, 129, 144
Leblanc, H. 144
Lemmon, E. J. 38, 63, 128, 129, 139, 144
Lewis, C. I. 126, 127, 128, 129, 144
Lewis, D. 129
Lindenbaum, A. 47, 50, 60

Makinson, D. vii, 63, 100, 101, 104, 105, 140, 144
Martin, R. vii
McKinsey, J. C. C. 129
Meyer, R. K. 63, 144
Montgomery, H. 127, 128, 144

Parry, W. T. 129
Pietarinen, J. viii
Pietczkowski, A. 44, 143
Poincaré, H. 3
Post, E. 129
Prawitz, D. 96, 144

Rasiowa, H. 115, 144
Routley, R. 63, 127, 128, 144

Salomaa, A. viii
Scott, D. S. vii, 96, 97, 98, 126, 139, 144
Scroggs, S. J. 129
Segerberg, K. 63, 140, 144
Sheffer, H. M. 128
Sibelius, P. vii
Sikorski, R. 115, 144
Stalnaker, R. 129

Takeuti, G. 96, 144
Talja, J. vii
Tarski, A. vii, 139, 140, 143, 144
Thomason, R. H. 63, 144
Tukey, J. W. 47, 49, 50

von Wright, G. H. 128, 144

Zorn, M. 48, 49